D0056244

PRAISE *for* *THE FUTURE OF HAPPINESS*

"Volumes have been written about how bad technology can be for us, but Amy Blankson is the first author I've seen to lay out a simple, achievable path to explain how to stay grounded and balanced in the digital era."

—Laszlo Bock, former head of people operations at Google and bestselling author of *Work Rules!*

"Too often, the concept of personal happiness is left out of discussions about technology and the future of our world. *The Future of Happiness* is a timely reminder about the importance of happiness, meaning, and our fragile, promising selves."

—Susan Cain, author of *Quiet: The Power of Introverts in a World That Can't Stop Talking*

"*The Future of Happiness* gave me tactical tips in the first few minutes and a genuinely happier life by the end. In an age of endless interruptions, this book couldn't have arrived at a better time."

—Neil Pasricha, *New York Times* bestselling author of *The Happiness Equation and The Book of Awesome*

"I believe that virtually all of us on this planet are struggling with two core issues: how to feel happier every day and how to balance technology so we're not overrun or dehumanized by it. Rather than fighting against the digital movement, we need higher-level help to harness it for our greatest joy, success, connection, and fulfillment. Amy's book is the perfect pathway to this—offering amazing new strategies to help us access more joy and meaning while at the same time leveraging the power of technology to enhance and enrich our lives. Thank you, Amy!"

—Kathy Caprino, MA, "Brave Up" writer, speaker, coach, and leadership developer

"Sometimes innovation doesn't have to be a brand-new technology, but a segment clearly redefined that can lead to greater well-being. I am so happy to see Amy address the link between technology and happiness. This book is long overdue for both professionals and parents, and I'm sure will be wonderfully received globally."

—John Stix, founder of KidsWifi

THE FUTURE *of* HAPPINESS

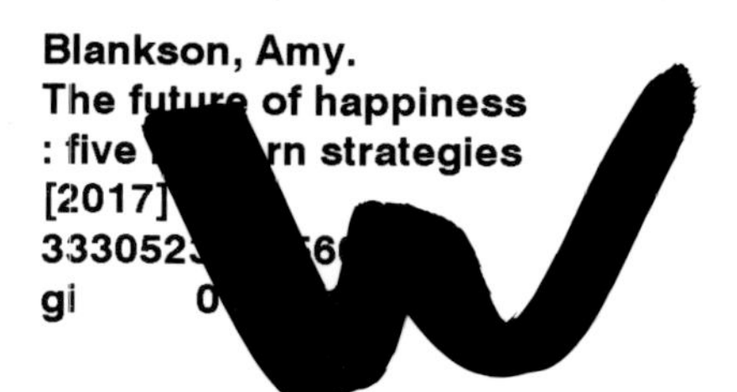
Blankson, Amy.
The future of happiness
: five modern strategies
[2017]
3330523
gi 0

THE FUTURE *of* HAPPINESS

FIVE MODERN STRATEGIES *for* BALANCING PRODUCTIVITY *and* WELL-BEING *in the* DIGITAL ERA

AMY BLANKSON

BenBella Books, Inc.
Dallas, TX

Copyright © 2017 by Amy Blankson

All rights reserved. No part of this book may be used or reproduced in any manner whatsoever without written permission except in the case of brief quotations embodied in critical articles or reviews.

BENBELLA

BenBella Books, Inc.
10440 N. Central Expressway, Suite 800
Dallas, TX 75231
www.benbellabooks.com
Send feedback to feedback@benbellabooks.com

Printed in the United States of America
10 9 8 7 6 5 4 3 2 1

Library of Congress Cataloging-in-Publication Data:
Names: Blankson, Amy.
Title: The future of happiness : 5 modern strategies for balancing productivity and well-being in the digital era / Amy Blankson ; foreword by Shawn Achor.
Description: Dallas : BenBella Books, 2017. | Includes bibliographical references and index.
Identifiers: LCCN 2016050073 (print) | LCCN 2017001179 (ebook) | ISBN 9781942952947 (hardback) | ISBN 9781942952954 (electronic)
Subjects: LCSH: Self-actualization (Psychology) | Motivation (Psychology) | Happiness. | BISAC: SELF-HELP / Personal Growth / Happiness. | BUSINESS & ECONOMICS / Motivational. | TECHNOLOGY & ENGINEERING / Electronics / Digital.
Classification: LCC BF637.S4 B566 2017 (print) | LCC BF637.S4 (ebook) | DDC 158.1--dc23
LC record available at https://lccn.loc.gov/2016050073

Editing by Debbie Harmsen
Copyediting by Brian J. Buchanan
Proofreading by Lisa Story and Michael Fedison
Indexing by Jigsaw Information
Cover design by Pete Garceau
Jacket design by Sarah Dombrowsky
Text design and composition by Aaron Edmiston
Printed by Lake Book Manufacturing

Distributed by Perseus Distribution
www.perseusdistribution.com

To place orders through Perseus Distribution:
Tel: (800) 343-4499
Fax: (800) 351-5073
E-mail: orderentry@perseusbooks.com

Special discounts for bulk sales (minimum of 25 copies) are available. Please contact Aida Herrera at aida@benbellabooks.com.

This book is dedicated to my three daughters, Christiana, Gabriella, and Kobi Lyn. May your futures be as bright as your smiles, and may your smiles brighten the future for all.

CONTENTS

FOREWORD

During the past ten years, I traveled to more than fifty countries trying to understand the connection between happiness and human potential. In that time, I have witnessed the rise of a positive movement emerging in our companies and schools. But I have also observed how much people are struggling. Depression rates are unacceptably high, isolation has become cancerous, knee-jerk medication is rampant, eating disorders and bullying are infecting our schools, and divorce and suicide are ubiquitous. These trends can and must be reversed if we are to find a society that flourishes.

But just when our need is greatest, we seem to have lost the most important of human faculties: the abilities to calm the brain, to focus our attention, to scan for the good, to envision a better world, and to prioritize connection. As a researcher who observes trends, I believe we must ask a root question as we start this book: *Does* happiness have a future? I believe Amy Blankson has the most researched and most well-thought-out answer I've seen to that question and to one of the largest challenges in the modern world.

We are living through twin revolutions. There's been a technological revolution that everyone knows about, allowing us to have smartphones and smartwatches. But hidden behind that technological revolution is a human one. Owing in part to that technology, we have been able to peer behind the curtain to understand what the human brain is doing as we construct our picture of reality. This has led to significant advances in the fields of positive psychology, neuroscience, and biofeedback. By combining these three fields with the technological revolution, we are beginning to see the most high-definition picture of human potential in history.

We now realize that (1) scientifically, happiness is a choice based on how we allocate our mental resources; (2) we can observe happiness spreading through social networks; and (3) we can now quantify how happiness is an incredible advantage in every domain of life. And technology can help us on that journey as we grow toward our true potential.

For those of you who don't know, Amy Blankson is my sister, and more famously, the "unicorn" from my TED Talk. My sister and I have been best friends since she was born. We played T-ball together. We were state debate champions in Texas together. We attended Harvard together. When she got married to the famed Dr. Bobo and had kids, I moved from Cambridge down to Texas to live on the same street as her family. So when I wanted to start a company to bring happiness research out to other people, I of course partnered with one of my best friends (and closest neighbors).

Over the past ten years, Amy has helped lead a company that has now worked with almost half of the Fortune 100 companies. She and I have traveled everywhere together: from London to work with architects to build happier cities, to NASA to find ways of using happiness to help propel us to Mars, to brain-trauma centers to research how virtual reality can help soldiers returning from combat.

One of my favorite memories is of Oprah Winfrey, sitting in her backyard in Montecito, California, calling in a singsong voice through the microphones on the *Super Soul Sunday* set for "Amy the Unicorn" to come join her in her garden just to give Amy a hug. She then invited Amy for a private lunch at her house to talk about the work she's doing in Africa. When you meet Amy, and I hope you will, you will see what makes her special. It's the rare combination of kindness and positive energy paired with experience and wisdom.

Amy is one of those people whose charisma is due to her humility. Since she's modest, I will brag on her. There is no one better to write this book than Amy. With her degrees from Harvard and Yale, her compassion and experience as the executive director of a nonprofit, with her understanding as the loving mother of three, and with her knowledge as a co-founder of the most well-known positive psychology consulting company in the world—Amy has her fingers on the pulse of what YOU can do in the midst of the busyness of life to harness technology to serve our true needs: social connection, meaning, and well-being.

What I like most about this book is that it avoids the pitfalls of condemning technology or becoming overly infatuated with it. Amy makes a fantastic research case about how we can lead, work, and parent during the largest disruption in human history. By keeping happiness, meaning, and joy as the true North, she provides a much clearer calculus for which technology to use, where to use it, and when to find safeguards against it.

I thought I'd be proud writing a foreword for my sister. I am, but that is not my overriding emotion. It is excitement. Excitement for the potential of this book to transform you and your family. Excitement for the massive companies and intrepid schools that have Amy helping them create well-being. Excitement for how small words in a Word document will eventuate in massive societal changes.

I am unapologetically biased. But I can also say with confidence that Amy has become the world's leading expert on the connection between happiness and technology. And she will be standing shoulder to shoulder with you at the vanguard of this movement as we try to help ensure that happiness has a long and bright future.

—Shawn Achor
Happiness researcher and *New York Times* bestselling
author of *Before Happiness* and *The Happiness Advantage*

INTRODUCTION

THE MODERN JUGGLING ACT

Technology is the biggest disrupter of happiness in human history.

This is the conclusion so many have come to in our world. In fact, in 2005, the *MIT Tech Review*, one of the centers of the technological revolution, posed a question: "Is technological progress merely a treadmill, and if so, would we be happier if we stepped off of it?" The author pointed to the Amish, who do without modern technology like cars and the internet, explaining that their depression rates are low, their happiness levels are high, and their life satisfaction rivals that of the top members of the Forbes 400.[1]

I travel around the globe speaking about happiness and technology. In fact, as I write this, I have just finished speaking to a group of 150 entrepreneurs in Munich, Germany. Before I began my talk, I was chatting with a few of the entrepreneurs about recent trends, and one of them leaned over to me and said, "Social media and technology are destroying our happiness, right?" I love the "right" at the end as if this is the only conclusion. But it is not. Disruptions only destroy when we cling to patterns that are failing.

While it's tempting to think nostalgically about life before the smartphone, the truth is that teens still ignored their parents and got into trouble back then. Kids still watched too much television and did not eat enough vegetables. Business leaders still felt as if they needed more than twenty-four hours in a day. Parents still struggled to balance work and life. We've had these problems for a long time, but the ways that we address these challenges today are different. We have too many tools—now many of them digital—so many that it's nearly impossible to decipher which ones to use effectively. They range from expensive to homemade, shiny to boring, sleek to cumbersome. And those characteristics don't define their utility.

I am a happiness researcher who cries when I laugh too hard. I'm an introverted mother of three who thrives in the midst of the pure chaos of a bouncy house birthday party with kids high on cake. I'm an impressionistic artist who loves data and spreadsheets. I'm a spontaneous individual who can also be very type A (my family jokes about how I used to save exactly 3 percent of my weekly $1 allowance to donate to save the whales). I'm a technophile and a nature lover, a romantic and a pragmatist. Some of my happiest moments with my three beautiful daughters are ones I can look back at over and over because I had the brilliant presence of mind to capture them on my phone. I have also missed out on meaningful moments with my family because I was on my phone. Within each of us are beautiful contradictions that bring complexity to how we perceive and address the challenges in our lives.

> Knowing that technology is here to stay, we need to learn how to live *with* the complexity of technology, not escape from it.

Knowing that technology is here to stay, we need to learn how to live *with* the complexity of technology, not escape from it. We need a strategy for using technology in a way that evokes our potential rather than hinders it. This book is about finding a balanced calculus that works in the modern world for deciding how to integrate the steady stream of technology into our lives in a way that fuels our success and happiness, instead of serving as a stumbling block.

The modern world is a juggling act, and sometimes it can seem like an all-out circus. A book that has a simple answer to happiness and technology is not one that works in a complicated world. And sometimes looking ahead seems like a luxury. But it's suddenly a luxury we can no longer afford because, by waiting for the world to come to us, we find ourselves one step behind our children and their technology. Without consciously curating which tech is actually best for our lives and businesses, we set ourselves up for frustration, outstripping our ability to enjoy any benefits the new innovations bring. What we need is a strategy to think ahead about where we really want to be going and how to create that future with what we have now.

Without consciously curating which tech is actually best for our lives and businesses, we set ourselves up for frustration, outstripping our ability to enjoy any benefits the new innovations bring.

A STRATEGY OF EVOKED POTENTIAL

I started the journey toward this book about thirty years ago, although I didn't know it. My dad, a neuroscientist at Baylor University, spent years studying perception and evoked potential (how the brain reacts to various stimuli). Despite the fact that he hooked my brother and me up to electrodes more than once to study our brain waves (I don't remember signing a consent form, Dad!), my father was the one who was most shocked (pun intended) that Shawn and I joined forces to create GoodThink, to bring the science of positive psychology to life for others. When we started the company in 2007, its focus was to help individuals find happiness in uncertain times using research-backed principles and strategies for sustaining a positive mindset. We traveled to over fifty countries, sharing this research with anyone willing to hear our message; we spoke to farmers in Zimbabwe, schoolchildren in Soweto, South Africa; bankers on Wall Street; and even leaders in the White House. However, gradually the questions we heard at our talks began to change. Instead of uncertainty about the economic health of the world, we began to hear concern about how technology is shaping our lives and those of future generations:

"Can happiness keep pace with innovation?"
"Would we be happier without tech?"

"How do we find happiness in spite of all this distraction?"
"How can we teach our kids appropriate tech boundaries?"

This book emerged from those real-life conversations with individuals across the globe, across economic boundaries, and across ages. I don't think I'm overstating it to say that the answers to these questions will define our time. These are the questions that undergird the modern family dynamic, that shape workplace efficiency and engagement, and that set the baseline for our interactions and communications with friends.

The future of happiness is up to us. By intentionally thinking about where, when, why, and how we are using technology, we can begin to actively shape the social scripts and market forces that drive our culture to create the future that we truly want to see.

A VISION OF THE BALANCED LIFE

In the last fifteen years alone, the number of patents granted by the US Patent and Trade Office doubled to more than 326,000 innovations being introduced every year.[2] Technology is flooding into every crevice of our lives faster than we can currently assimilate and evaluate what this new technology means for us. We struggle to know which advances are best for us, especially when reports are released every day that change the script of how we are supposed to engage with technology. For example, we are now told:

- ✓ Place your devices at eye level so that your neck is straight and neutral.
- ✓ Take a twenty-second break every twenty minutes to look at something at least twenty feet away (the 20/20/20 rule).
- ✓ Stand up and move around every 1.5 hours.[3]
- ✓ During these short breaks, don't get distracted because it will take you eleven minutes to refocus and potentially double your rate of error on tasks.

As I write this, I hear Charlie Brown's teacher in my head . . . womp, womp, wa-womp, womp. In the course of writing this book on balancing productivity and well-being in the Digital Era, I've spent hours reading about how tech is killing me softly—leading to spinal deterioration, headaches, insomnia, blurry vision, carpal tunnel—and yet, I still have no intention of removing technology from my life. During this book process alone, I logged more than 500 hours on the computer, spent eighty-plus hours on my phone researching apps, downloaded and tested several dozen new apps—and visited the chiropractor six times for back and neck pain. The irony is not lost on me. I've got all this data and info . . . and yet I continue to delude myself that I will be the one exception to the rule. I might be slouching now, but I'll go for a run later and it will all even out, right? Right?!

The truth is that I have developed selective hearing (my husband has suspected this for years). I know technology can be both helpful and harmful, but I'm so tired of hearing only one side of the story that I often tune that side out. What I want is a thoughtful, measured vision of what a balanced life could look like in the future, and then I want to know how to emulate that.

That, my friends, is why I've devoted the past few years of my life to researching how to balance productivity and well-being in the Digital Age. It's why I have personally tested more fitness wearables, journaling apps, meditation timers, and posture-training devices than I can count. It's also why I have spent countless hours collaborating with companies on the "digital frontier" like Google, Plasticity Labs, the Center for Brain Health, and Happify that are using technology to improve well-being. Through these experiences, I've gained some valuable insight and thought-provoking stories that I can't wait to share with you. In Part One of this book, I'll address the top three burning questions that I hear most often about the future of happiness; and in Part Two, I will outline five strategies for maximizing your productivity and well-being that I have collected from experts and innovators all over the world. My goal is to inspire through this book a new way of thinking, one that gives us a language with which to think about technology and happiness and a vision for how to balance the two effectively in the future.

THIS BOOK IS NOT JUST FOR TECHIES

It's worth saying that this book is not just for techies. It's for anyone who struggles with the growing number of charging devices taking over his nightstand. It's for executives and employees who are striving for increased productivity but find their eyes burning and backs hurting after hours of staring at their screen. It's for the minivan moms who love how devices help to calm their children while in the car or in public, but then have to deal with "tech tantrums" when the device is turned off. It's for young professionals overwhelmed by their digital footprint and needing to strategically unplug.

If you would like to read this book as it unfolded in my head, I encourage you to read it from start to finish. However, if you have a particular area of interest, feel free to scan the Table of Contents and pick the strategy that speaks to you the most. Are you having trouble staying grounded? Are you frustrated with your work or home spaces being overrun with technology? Are you looking for fun, new technology to accomplish your fitness goals, improve your posture, or manage your time better? Whatever your personal interests, this book is aimed at helping you raise your consciousness about tech in your life and become more intentional about how you fuse (or lose) tech in your life to maximize productivity and well-being.

> Throughout the book, you will see sections entitled "Happy Hacks to Get You Started" with practical tips for how you can start putting these ideas into action.

PART ONE

THE THREE BURNING QUESTIONS OF THE DIGITAL ERA

WHERE ARE WE HEADING?

LURED BY THE LIGHTS

For thousands of years, baby loggerhead sea turtles have relied on the light of the moon and stars reflecting on the sea to find their way to the ocean immediately after hatching.[1] This amazing process worked perfectly until beachfront developments along the coast in Florida began to change the landscape and lighting of the natural habitat. Disoriented by artificial lights, more than half of the baby turtles began to crawl toward the city within the first few minutes of their lives—straight toward the highway and other natural predators, creating a deadly trail of hundreds of endangered animals every night. In response, the City of Fort Lauderdale developed one of the strictest lighting ordinances in Florida, calling for an artificial-lighting curfew at 9 PM, literally blacking out 45,000 feet of beachfront property.[2] However, the community pushed back both for safety reasons and for tourism. An article in the *Sun Sentinel* asked, "Which should get priority? A city neighborhood literally left in the dark or an endangered species harmed by street lights?" Tensions ran high until 2011, when turtle-friendly lights were introduced to the market. A cost-effective solution, turtle-friendly lights utilize an amber-colored light with a unique light-wave not visible to turtles. Since that time, the rate of hatchling disorientation in Fort Lauderdale has dropped from 50 percent to zero.[3]

DISORIENTED BY DISTRACTION

Like the sea turtles, we are drawn to the bright lights of our phones, tablets, laptops, and TVs, our minds and bodies becoming disoriented as we lose focus and direction. Each day, we are torn between the value of tech and the cost to our health. There's no denying that we are in the midst of an incredible digital revolution that is rapidly morphing the landscape of daily life with new time-saving tech tools for productivity, efficiency, and communication. These tools *should* be fueling our happiness levels, but that doesn't always seem to be the case. In fact, the timing of the digital revolution coincides with record-setting levels of depression and overall life dissatisfaction.

Since the advent of the personal computer in the 1970s, depression rates in America have increased tenfold.[4] And over the past three decades, obesity rates in the adolescent population in particular have *quadrupled* (Centers for Disease Control, 2012). In my husband's work as an adolescent-medicine specialist, he has seen this trend firsthand, as device-wielding tweens and teens strut into his office with little insight into the inverse relationship between their decreasing exercise and increasing screen time. Although tech offers many advantages in our lives, its deleterious impact on our health and happiness is not fully understood. We have spent 99.9 percent of human existence without high tech, and now that it's here we have a lot of catching up to do from a biophysical/emotional standpoint. There has been a proliferation of new research on tech anxiety and social disconnection that warrants our taking a step back to think hard about where we are going in the future. The younger generation spends an average of six-plus hours per day on their phones, literally changing a quarter of their life experiences from what we have known in the past.

As devices become smaller and more sophisticated, we are taking them into the bedroom and disrupting a basic human function—sleep. Clinical psychologist Amanda Gamble explains that many people have begun taking their devices to bed—a phone as an alarm clock, a tablet as a book, or a laptop as a television—which deprives the brain of the downtime that's necessary in sleep.[5] Like sea turtles, our brains are getting confused by the artificial screen lights, signaling our brains to stay awake, thinking that light means it is daytime, and creating a break in our natural circadian rhythms. As a result, our brains produce less of the hormone melatonin, which helps regulate sleep and further entrenches the problem. The problem especially poses a danger for still-developing children and teens, as this learned association puts them at much greater risk of developing mental health disorders like anxiety, depression, and addiction to substances, as well as physical problems like poor glycemic control, diabetes, and insomnia.

> With 95 percent of Americans now spending two or more hours each day using a personal digital device, a new host of medical complications is just now beginning to emerge.

With 95 percent of Americans now spending two or more hours each day using a personal digital device, a new host of medical complications is just now beginning to emerge. "Texting thumb" is a new repetitive-stress injury, like carpal tunnel syndrome, attributed to texting and video games. In addition, 70 percent of millennials report symptoms of digital eyestrain (that's more than baby boomers at 57 percent and gen-Xers at 63 percent). Doctors are also reporting increased spinal and muscular tension due to what has been termed "text neck," which refers to an overuse syndrome or a repetitive-stress injury in which the head hangs forward and down while the person looks at mobile electronic devices for extended periods.[6] A study led by Erik Pepper of San Francisco State University showed that 84 percent of subjects reported some hand and neck pain during texting.[7] Moreover, subjects also displayed other signs of tension, like holding their breath and increased heart rates. Dr. Kenneth Hansraj, chief of spine surgery at New York Spine Surgery & Rehabilitative Medicine, explains that when the spine is in a neutral position, the head weighs about ten to twelve pounds. However, when the head leans forward just fifteen degrees, the neck feels like it is supporting the equivalent of twenty-seven pounds; when the head bends down forty-five degrees, the neck feels forty-nine pounds; and at sixty degrees, it feels sixty pounds. As Dr. Hansraj expounds, "That's sixty pounds of weight stress on muscles and nerves that are meant to handle ten to twelve pounds of stress, and that much load can do a lot of damage over time."[8] The cervical discs take on this additional load, causing premature disc degeneration twenty or thirty years earlier. Concurring, Dr. James Carter, a former governor of the Australian Spinal Research Foundation, said "text neck" can lead to a four-centimeter spinal shift with repeated head tilts, a particularly sobering statistic given that he has seen a 50 percent increase in the number of patients who are school-age teenagers.

The implications of our physical bodies' morphing in response to external devices are far-reaching. Harvard professor Amy Cuddy even found that hunching over your phone can foil your workplace confidence—not only does slouching hurt your back, but the posture also hurts your ability to speak up and be assertive.[9] At this rate, imagine what the next iteration of the famous "evolution of man" illustration will look like in fifty more years—no longer will we be a species walking straight and tall, but rather we will be curved and hunched human forms gripping a cellphone with a grimace on our face. Is this who we want to become?

The modern evolution of man by Kevin Renes/ShutterStock.com.[10]

CHOOSING OUR ADVENTURE

Technology intersects with our lives constantly, shaping not only our health but also our livelihood. Despite wonderful intentions, in many ways the technological revolution has left us more fragmented and insecure.

Software and IT experts have never been busier as they seek to find better ways of processing high volumes of data. Millennials, a group that should be comfortable with this emerging tech, often find themselves in digital overload. And parents feel exasperated trying to keep track of their children's online activity, while stressing about their own online identity and privacy. We have amassed an overabundance of data without knowing how to use it or even where it all goes. The data in many ways is disconnected from our daily lives, and when it does emerge, often it's as a nasty surprise in the form of identity theft or lawsuits.

I don't know about you, but sometimes I feel saddened by our reliance on technology. Nothing makes that sentiment more real than sitting in a restaurant and looking around at "families" having dinner together. I use "together" loosely because it really appears that four or five individuals are having personal dinners at the same table, each person zoned out on a hand-held electronic device. One recent story I heard from an NPR host was that his child drew a family picture and in his hand there was an iPhone. Another woman at a conference I was at joked that her son didn't even draw her face in the family picture—he drew her as a laptop. It was her wake-up moment. When I give talks around the country, I see employees in the back row of the conference hall entranced by the shiny screens of their laptops or cellphones (maybe they are all taking notes of me speaking about happiness? I'm an optimist). As tech advances and we accept these changes without pause, I worry that maybe our happiness is getting left behind, moving further down the priority list.

Technology has come barreling into our lives like a wild stallion in recent years, leaving us with two choices: jump on and hold on for dear life, or let the stallion gallop past (until we realize there's a lasso attached to our ankles and we are dragged along anyway). For those of us who jumped onto the horse, we are just beginning to realize that we have no idea where the stallion is going . . . and this bareback riding has us exhausted. We are at a crossroads—and in desperate need of a saddle and bridle. How can happiness keep pace with technological development in our lives? As much as we love the novelty and excitement of tech, it's time for us to take control so that we can harness the speed and power underneath us. For those of us being dragged along, maybe now we wish we were "back in the saddle with a whip and stirrups," anything to provide some semblance of control since Techbiscuit shows no signs of slowing down. It's time to choose our adventure: Are we going to succumb to this wild ride as overwhelmed consumers? Or are we going to take charge of the future by becoming co-creators in the way that new inventions intersect with our work, our families, and our communities?

As much as we love the novelty and excitement of tech, it's time for us to take control so that we can harness the speed and power underneath us.

WOULD WE BE BETTER WITHOUT TECH?

BEFORE PANDORA'S BOX

Remember "the good ol' days" when we used to make calls on telephones attached to walls? I recently asked my eight-year-old what a dial tone was and she just stared at me blankly. Now almost everyone (and their mother and possibly their grandmother) has a smartphone. And they expect to be able to reach you at all times. If you don't pick your phone up, they can and will use other means of communication: voice mail, text, email, calling people who might be near you. And if all else fails, and they have this ability because you share an account with them (disdainful throat-clearing), they may ping you as if you were a lost homing beacon (I do love my husband's concern for my well-being and his need for an immediate answer to that oh-so-important question of "leftovers or meet out for dinner?")—which will alert everyone around that you have (a) been trying to avoid someone, or (b) incompetently left your phone on vibrate and misplaced it.

I'm not a particularly superstitious person, but every time I leave the house, a little voice in my head pipes up, "Don't forget your phone. If you leave it, something will happen."

Of course, that one time I left my phone at home, my car was struck by a hit-and-run driver, and I had no way to report the incident. When I got home to call the police, they chided me for not staying put and filing a police report right then. I cringed, knowing Murphy's Law was at work. I'm not a particularly superstitious person, but every time I leave the house, a little voice in my head pipes up, "Don't forget your phone. If you leave it, *something will happen.*" Researcher, speaker, and author Brené Brown would call this our brain's way of "dress-rehearsing tragedy," or steeling ourselves against potential negative outcomes by jumping ahead to name the worst possible outcomes. Somehow we think if we can say the outcome first, we will avoid getting jinxed. But the research shows that this behavior is actually maladaptive, reinforcing a kind of learned helplessness that only makes the outcome that much more likely. And we wonder why we have anxiety about tech in our lives . . . In this uber-connected vortex we live in, it is essential that we consciously utilize technology so that we don't become codependent, afraid to take a breath without an electronic pulse beside us.

In this uber-connected vortex we live in, it is essential that we consciously utilize technology so that we don't become codependent, afraid to take a breath without an electronic pulse beside us.

ROBOHAPPINESS

Technology has moved away from merely making our lives more convenient. It now has the potential to change every aspect of what we are as humans, from how we pursue happiness to how long we live to how we connect.

Enter transhumanism. A fancy word to describe the plotlines of some of pop culture's best movies—*The Terminator*, *The Matrix*, *Ex Machina*, *Avatar*. Transhumanism is a movement that aims to explore how technological innovation can help us surpass our natural limitations, literally becoming superhuman. We are on the cusp of these futuristic movies becoming our reality, an exciting and simultaneously frightening development.

Ever dream of being Tom Cruise in *Minority Report*, able to swipe through information in the air? Get ready for Project Aura, a pair of glasses designed by Google not only to "offload" memories for later recall, but also to allow users to communicate in foreign languages and to view meaningful information about their surroundings in their peripheral vision.[1,2]

Did you ever think that *Robocop* or *Iron Man* would become a reality? Lockheed Martin is currently testing the Human Universal Load Carrier (HULC), an exoskeleton for soldiers to wear that can help a soldier carry up to 200 pounds while running at ten miles an hour for extended periods. Video games like *Call of Duty* already allow gamers to virtually experience this type of technology. Perhaps you thought that the movie *AI* (*Artificial Intelligence*) was just a sci-fi fantasy dream? Think again. Ray Kurzweil, the director of engineering at Google, recently predicted that by the 2030s, humans will become hybrids, meaning that we will be able to use technology to merge and enhance ourselves to transcend our natural limitations.[3] He predicts that our brains will be able to connect directly to the cloud via nanobots (tiny robots made from DNA strands) to augment our intelligence. What's even scarier? Of the 147 predictions Kurzweil made in the 1990s for the year 2009, 86 percent were correct. He writes that "technology is a double-edged sword. Fire kept us warm and cooked our food but also burnt down our houses. Every technology has had its promise and peril."

In the same way that we anticipate what might be inside a gift box, there is a giddiness and excitement about opening Pandora's box. As we tear off the pretty wrapping paper and behold the shiny, new object inside, we are enthralled and obsessed at first. The excitement always overshadows the potential pitfalls. But newness eventually fades, and we begin to see that the contents of Pandora's box come with inherent risk. Just as in the transhumanist movies of pop culture, there always seems to be an ugly side of futuristic technology wherein the machines almost take over. That should scare us. Or warn us. Or cause us at least to pause to be intentional about where we are heading.

We are approaching the precipice of being able to engineer our own evolution through nootropics (pills for cognitive enhancement), genetic manipulation, and nanomedicine.[4] It's hard not to get excited about the possibilities of becoming smarter, stronger, and healthier—who among us wants to see loved ones get sick or suffer or even forget who they are? Suddenly, becoming *super*human seems possible for the first time in the history of humanity . . . but what does this mean for society?[5] And will advances like this actually make us any happier?

Oxford scholar and senior fellow professor Nayef Al-Rodhan explains that humans are genetically and neurochemically hardwired to want to feel good.[6] In his article *Inevitable Transhumanism? How Emerging Strategic Technologies Will Affect the Future of Humanity*, Al-Rodhan writes that five key factors drive our behavior: power, profit, pleasure, pride, and permanency. Any technology that enhances one of these five factors is likely to be adopted, since it will appeal to the feelings that make us "feel good." This drive pushes us further and further to a transhumanist outcome, where "the human experience is artificially enhanced or changed . . . It's not 'how' or 'if,' but rather 'when' and 'at what cost.' "

While Al-Rodhan paints a bleak, if not outright frightening, picture of the future, not all transhumanist outcomes are threatening to our nature. In fact, some of these outcomes are downright miraculous. For instance, thirty years ago, Hugh Herr was hailed as one of the best rock climbers in America. However, at age seventeen, when Herr and a friend embarked on an ice climb on Mount Washington, a blizzard engulfed the climbers and caused them to stray into a remote chasm. After three days of wandering without snowshoes, the two were miraculously rescued, though suffering from severe hypothermia. In the days that followed, both of Herr's legs were amputated below the knees, which was a devastating blow for Herr both personally and professionally as a rock climber. However, in the months after the accident, Herr set to work crafting his own artificial legs out of rubber, metal, plastic, and wood in a local machine shop. Within five months, to the amazement of his peers, he returned to Mount Washington and resumed his climbing practice. Today, Herr is an MIT faculty member and director of the biomechatronics research group, where he designs smart prostheses for maimed athletes.[7] Herr is known to joke around the office that while everyone else is growing older, his legs are getting younger.

Yet, as our desire for self-preservation pushes us toward new technological solutions, many are concerned about the physical and spiritual implications of becoming hybrid humans with metallic implants and 3D-printed organs that lengthen life. What are the environmental and personal impacts of living longer? What effect would transhumanism have on our humanity? As Rafael Calvo writes, "If a technology doesn't improve the well-being of individuals, society, or the planet, should it exist?"[8] Whether you are a business owner, a researcher, an engineer, a government official, or a citizen, we all have a responsibility to help answer that question and to shape the future of happiness.

AUGMENTING THE MIND

In his recent TED Talk, tech innovator Meron Gribetz explains, "Today's computers are so amazing that we fail to notice how terrible they really are." He goes on to recall a moment in 2011 when he visited a bar with a friend but couldn't maintain a meaningful conversation because his friend kept receiving and responding to text messages. Frustrated, Gribetz looked around him and saw across the room a group of teenagers huddled around a phone, engaged and laughing together over Instagram photos.[9] He said that "the more I thought of it, the more I realized it was clearly not the digital information that was the bad guy here, it was simply the display position that was separating me from my friend and that was binding those kids together. See, they were connected around something, just like our ancestors who evolved their social cognitions telling stories around the campfire. And that's exactly what tools should do, I think. They should extend our bodies. And I think computers today are doing quite the opposite.

Whether you're sending an email to your wife or you're composing a symphony or just consoling a friend, you're doing it in pretty much the same way. You're hunched over these rectangles, fumbling with buttons and menus and more rectangles. And I think this is the wrong way, I think we can start using a much more natural machine. We should use machines that bring our work back into the world. We should use machines that use the principles of neuroscience to extend our senses versus going against them." Meron Gribetz went on to become the founder and CEO of Meta, a company that designs an augmented-reality experience to extend our humanity, not hinder it through the use of glasses (imagine a 3D version of FaceTime with the capability to share digital objects across space and time—mind-blowing!).

Unlike virtual reality that uses artificial intelligence to create an imaginary world, augmented reality is designed to aid the human mind in processing the real world. Augmented reality focuses on the human brain as the ultimate processor and enables you to boost not just the quantity of information, but also the quality of information in your environment. Devices (wearables, ingestibles, embeddables) may be intertwined with the natural capabilities of our human body and mind, but the goal is to make these devices so transparent that we do not even notice them while they improve the quality of our air, the acuity of our vision, the productivity in our work, and, most important, the breadth and depth of the joy we feel striving after our potential, our happiness.

We need to find a way not just to coexist with technology, but to thrive with it.

WHAT WILL HAPPINESS LOOK LIKE?

The ancient Greeks defined happiness as "the joy we feel striving after our potential."[1] Although I love this definition (and use it as the basis for my understanding of happiness throughout this book), much has changed in the world since ancient Greece. As a result, the strategies that we use to pursue happiness require a more nuanced perspective. While there are many happiness books on the market today, few of them look at happiness in the context of the unique pressures and demands of the technological world that we live in today. This book is the first to apply the principles and research from the field of positive psychology to help us find a sense of balance between productivity and well-being in the Digital Era.

Everyone warned me as I was writing this book . . . *Amy, don't get all philosophical—keep it light, keep it practical, or you will scare everyone away.* But you know what? I'm going there anyway, because like a kitten unraveling a ball of yarn, I find it fun to chase a thought to the very end. And besides, how many of us haven't wondered whether life wasn't better in the good old days, before we opened Pandora's box? Now that we have opened it, is there any way to stuff the technological revolution back into the box and close it? Not likely. But depending on which version of the Greek myth you've heard, you may have forgotten that when Pandora first opened the box and released evil and pestilence into the world, one thing remained at the bottom of the box: hope. I truly believe that technology, despite its potential to bring out the "evils" within us, also has the potential to save us. My desire is that we turn that box of tech upside down and shake loose enough pieces of hope and happiness to improve our future.

STRATEGIES OF THE HAPPIEST PEOPLE IN THE DIGITAL ERA

Before I began writing this book, if you had asked me to make a character sketch of someone I thought would have the greatest sense of well-being in the Digital Era, I would have described someone who lived an ascetic lifestyle—a life of self-discipline and denial of worldly pleasures—far off in some remote location, meditating by day and sleeping unhindered by alert notifications by night. I couldn't have been more wrong.

I have since had the opportunity to interview countless individuals from around the world—from the heart of Silicon Valley to the backwoods of Smallville USA, from high-powered executives who manage employees around the clock to stay-at-home moms who rely on electronics to make it through the day, from children born with an iPhone in their hands to individuals in their sunset years who have just learned how to use a mouse—and what I learned was this: the individuals who reported the greatest sense of balance and ultimately well-being could be any of these typologies. They might have vastly different experiences with technological devices, but ultimately the most balanced, satisfied, and happy individuals use five key strategies, not just to survive but also to actually *thrive* in the Digital Era:

- ✓ First, they stay grounded in the face of distraction.
- ✓ Second, they use technology to know themselves on a deeper level.
- ✓ Third, they know when and how to use technology to train their brains to reach their full potential.

- ✓ Fourth, they structure their surroundings to create a habitat for happiness.
- ✓ Fifth, they innovate consciously to enrich the world around them.

Over the course of this book, you will meet some of the pioneers in the tech industry working at the forefront of the field of digital well-being: individuals like Javier at the MIT Media Lab, who is studying "communication prosthetics" for individuals with autism; Dane at iMotion Labs, who is using facial analysis to help companies understand consumers; and Ofer at Happify, who has developed a framework to gamify cognitive brain-training. The journey of these pioneers is incredibly important in light of the growing need to understand well-being, productivity, and success both in research and in real life. And my hope is that the research and experiences presented in this book will inspire and challenge you to go deeper into your own thinking, to put these strategies for wiring your world for greater happiness into action, and to share the research behind these strategies with friends, colleagues, and even children to change the trajectory of when, where, why, and how we use technology to influence our well-being. Now is the time to embrace the spirit of opportunity and truly make a difference in the world. This is the defining moment in the future of happiness.

PART TWO

THE FIVE STRATEGIES FOR BALANCING PRODUCTIVITY AND WELL-BEING

STRATEGY #1
STAY GROUNDED

HOW TO FOCUS AND CHANNEL YOUR ENERGY WITH INTENTION

"You can see the computer age everywhere but in the productivity statistics."
—Robert Solow, Nobel laureate in economics

When I was six and my brother was eight, my dad came home from one of his neuroscience conferences with an educational kit for us on the science of electricity. Now mind you, this kit was from the 1980s so it did *not* dutifully include appropriate age guidelines (seeing as my kids are now six and eight, I can confidently say I would seriously hesitate before giving them a similar Pandora's box—I mean, science kit). Like most kids, my brother and I had no interest in reading the instruction booklet (tiny print, too many complete sentences). We simply dumped out the box of wires and gadgets and started playing. In fairly short order, we had lightbulbs blinking, sounds buzzing, and even balloons filling with air. But then my brother got the bright idea to try out one of the little alligator clips on our night-light. He snapped one clip onto a prong of the night-light and precariously plugged the unit into the wall. Nothing happened and we paused, momentarily puzzled.

In a last-ditch effort to make the night-light cooperate, Shawn attached the other side of the alligator to the other prong, and yep, you guessed it, we got a reaction when the circuit was completed. With a loud pop, my brother was blown back against our bunk beds, his hair standing straight on end. Our mother came running into the room after hearing the boom and demanded to know what was going on. "Nothing!" we said in unison and smiled, as smoke emanated from the blackened wall and swirled around our heads. Lesson learned.

I love that I have grown up around technology—it's fascinating and helpful and fun. It's also incredibly powerful, distracting, and even potentially dangerous. As Shawn and I learned in our experiment, a two-prong electrical cord can pull lots of power, but it can also start fires. In the coming section, I will explore how digital distraction is affecting our happiness and share ways that you can develop "a third prong" to help you ground your energy in a rapidly changing world.

THE CHALLENGE:

NAVIGATING DIGITAL DISTRACTION AND ADDICTION

If I were to ask you to give me a word associated with "tech addiction," you would likely say "teens" before I could even finish my sentence. And your stereotype would be well founded; in a 2016 survey of 1,200 teens and their parents by Common Sense media, 50 percent of teens admitted to feeling addicted to their mobile devices.[1] However, teens are not the only problem. In the same survey, 27 percent of parents also admitted to feeling addicted. And while 77 percent of the parents felt that their teens get distracted by devices and don't pay attention when they are together, 41 percent of teens say the same about their parents. Touché. The survey goes on to report that 48 percent of parents feel they have to answer emails and texts immediately, and 69 percent of parents say they check devices hourly. Even if you are not one of those people who would admit to being tech-addicted, I'm willing to bet that you know someone who struggles with being tied to a device, whether he or she wants to be or not.

In the course of writing this section, I remember walking into my office one day when something caught my eye. I turned my head and noticed my Christmas tree, sparkling in the corner of the room. "It's the middle of January!" I thought. "I really should take that down—it will just take a second." And then upon dismantling the tree, I realized that empty spot would be the perfect place in my office for my dusty treadmill. "This is a great time to jump-start my New Year's resolutions!" I was just about to begin measuring the space when I remembered that I was supposed to be focused on writing . . . about distraction. Ugh. My saving grace in writing this section is that I now know from my research that I'm in good company with the millions of other adults who are distracted. In fact, adults are becoming increasingly distracted by our fast-paced world. Wait, where was I going with this? Oh, right. What's the greatest enemy of being productive and staying grounded? Distraction.

TO INFINITY AND BEYOND

Sometimes distraction is even built into our entertainment. For example, the Toy Story ride at Disney World is a perfect example of what it means to be distracted amid a fast-paced world. The ride places you inside a spinning car mounted with toy laser guns for each member of your party. As you move into the ride, shiny targets cover every inch of your visual field. Just as you begin to hone your aim, the ride swings the moving car in a new direction with fresh targets popping out around you. The challenge of the game is that in a spinning car with so many targets, it is virtually impossible to aim.

Tech, like this theme park ride, is visual candy—so bright, alluring, and fun that you almost forget the purpose of the ride is just to relax and enjoy. To deal with this dizzying array of targets, our brains have gone into autopilot, autofocus. The problem is that we don't have a vision for how to fuse technology into our lives, so we constantly feel like spinning pawns in an ever-changing game. We keep moving because we are on a trajectory that seems to be going somewhere fast, even if we don't know where. If we want to go "to infinity and beyond," we need to figure out how to interact with technology with our feet firmly planted on solid ground, our intentions serving as a guiding light in a sky filled with attention-grabbing supernovas.

THE IMPACT OF DISTRACTION ON FOCUS

In 2013, the National Center for Biotechnology Information reported that the average attention span of a human had dropped to a mere eight seconds (from twelve seconds in 2000); meanwhile, the average attention span of a goldfish is nine seconds. While this factoid may be shocking, chances are that you will move on to another thought and likely forget this information by the end of the next paragraph.

In 2013, the National Center for Biotechnology Information reported that the average attention span of a human had dropped to a mere eight seconds (from twelve seconds in 2000); meanwhile, the average attention span of a goldfish is nine seconds.

Why does this matter? According to Cyrus Foroughi, a doctoral student at George Mason University, one minute of distraction is more than enough to wipe your short-term memory.[2] Most interruptions in the real world can last anywhere from ten to fifteen minutes, a troublesome statistic given that an interruption as short as 2.8 seconds (the length of time it takes to read a short text message) can double error rates on simple sequencing tasks, and a 4.4-second interruption (like sending a text) can triple error rates, which is a serious problem when you consider texting and driving habits.[3] Our jobs today are "interrupt-driven," with distractions not just a plague on our work—sometimes they can mean the difference between success and failure.[4] Yet, in many professions like medicine or engineering, doubling or tripling errors can have life-threatening implications.

The science of interruptions began more than 100 years ago, when researchers switched from studying labor-saving devices like conveyor belts in the Industrial Revolution to mind-saving devices like data entry in the tech revolution. Psychologists began looking at how interruptions affected the work of telegraph operators, who had to rapidly deliver time-sensitive information, and found that when someone spoke to the operators, the operators made more errors because their brains had to "switch channels" between work and conversation. While interruptions were unavoidable, researchers learned that *how* the operators were interrupted was key. As NASA learned with its astronaut program, if interruptions were too distracting, they could throw off the astronauts who were doing experiments while also monitoring potentially fatal errors. If the interruptions were too unobtrusive, they might go unnoticed and cause more problems. So in 1989, NASA hired researcher Mary Czerwinski to devise "the perfect interruption" to communicate with busy astronauts in space. She knew that sounds could be too jarring, and she also learned that most of the messages that astronauts received were text and number based. To differentiate "the signal from the noise," she used a visual graphic whose sides changed color depending on the type of problem at hand, which turned out to be highly effective for communicating with minimal interruption.

Almost thirty years later, we are still looking for the perfect interruption, although this time the messages are coming through wearables—devices you wear—like Project Aura, Laster SeeThru, or Icis.[5,6] Using the inside of a pair of eyeglass lenses for a screen, messages can be displayed about the external environment on an ongoing basis—and they aren't necessarily provided as "the perfect interruption."[7] In other words, if you think tech is distracting now, brace yourself for a whole new level of distraction in the not-so-distant future when employers could theoretically and quite literally "get in your face" to relay urgent messages or provide status updates.

THE DISTRACTION EPIDEMIC

Linda Stone, a software executive who has worked for both Apple and Microsoft, explains that we are so busy keeping tabs on everything that we never focus on anything, a phenomenon she calls "continuous partial attention." If only we could have a personalized NASA control center guarding our concentration! Instead, messages undiscerningly bombard us, with the senders rationalizing that we can choose when and where to open a message. What initially began as a study in information efficiency has quickly turned into a global marketing quest to create the most attention-arresting devices known to man—alerts, beeps, taps, vibrations.

- ✓ 67 percent of cell owners find themselves checking their phone for messages, alerts, or calls—even when they don't notice their phone ringing or vibrating.[8]
- ✓ 44 percent of cell owners have slept with their phone next to their bed because they wanted to make sure they didn't miss any calls, text messages, or other updates during the night.
- ✓ 29 percent of cell owners describe their cellphone as something they "can't imagine living without."
- ✓ 55 percent of workers reported checking their email after 11 PM—and 6 percent reported checking email while they or their spouse were in labor![9]

Digital interruptions fall into a sort of Heisenbergian uncertainty trap: How can you know whether an email or text message is worth reading unless you open and read it?—which is an interruption in itself! Typical mobile users check their phones more than 150 times per day, and the average office worker checks their email thirty times every hour—that's every two minutes![10] As *New York Times Magazine*'s Clive Thompson writes, "Information is no longer a scarce resource—attention is."

Curious about the real-world implications of these distractions, Gloria Mark, a professor of informatics at the University of California, Irvine, set out to measure how high-tech devices affect our behavior. Beginning in 2004, she began studying employees at two high-tech firms. She persuaded her graduate student Victor Gonzalez to look over the shoulder of various employees for more than a thousand hours, noting the number of interruptions as well as the length of undisturbed focus. When Gloria crunched the data, the results were "far worse than I could ever have imagined." Each employee spent an average of only eleven minutes on any given project before moving on to the next task. Each time a worker was distracted, it would take an average of *twenty-five minutes to return to that task.*

Inspired by Professor Mark, I decided to run my own five-minute experiment and count the distractions in my life, and here's what I found: dog barked, phone rang, package was delivered, phone dinged for news update, random stranger stopped by to ask a question about Wi-Fi, my mom called—six interruptions in five minutes. The sum total effect of these distractions? I gave up trying to focus because I felt like my mom's call took precedence, and I would try to pick up where I left off . . . later. Interruptions leave us feeling desired and needed, which can become intoxicating and addictive. Thompson goes on to explain that, "The reason many interruptions seem impossible to ignore is that they are about relationships—someone, or something, is calling out to us. It is why we have such complex emotions about the chaos of the modern office, feeling alternately drained by its demands and exhilarated when we successfully surf the flood."

In response, some companies like Runcible have devised what they call the anti-smartphone, a spherical cellular device that tries to solve the program of notification overload by stripping down its functions to glanceable information that can save you from constantly pulling out your smartphone.[11] While you can still make calls and browse the web, there are no temptations to check social media or work emails because there are no apps on the phone.[12] CNN Money jokes that this device is a tempting alternative for those who feel that "simply unplugging or tossing a phone into a fire isn't always practical."[13] However, the article goes on to warn potential buyers that the device might have unintended side effects, like constant nervous checking for messages. It seems that distraction is perhaps more a state of mind than an external force.

THE PRODUCTIVITY PARADOX

For years, we have bought into the idea that technology is supposed to help us to become more productive, so that we can use more of our free time to do the things that make us happy. Yet as Robert Solow, Nobel laureate in economics, famously quipped in 1987, "You can see the computer age everywhere but in the productivity statistics."[14] As a society we have bought into the idea that smart tech can do smart things; but using smart tech *smartly* is a whole other issue. Companies are finding that just because they invest in and introduce new tech like data-record systems or project-management software to their workforce doesn't mean that the tech will be integrated or even appreciated.[15] For instance, in the medical field, digital-record systems are designed to reduce errors due to penmanship or misfiling, but if there is an internet glitch, the entire office can come to a grinding halt, leading to a reversal of the expected benefits. When new innovation doesn't jive with how people actually work on a day-to-day basis, tech can actually decrease productivity and happiness.[16]

As a society we have bought into the idea that smart tech can do smart things; but using smart tech *smartly* is a whole other issue.

THE HAPPINESS CLIFF

Sometimes tech is fun just for the sake of the endorphin rush and the dopamine boost. But at what point do those focus-altering diversions cause us to lose sight over what we really care about? At what point do diversions turn into fixations that are distracting?

Sometimes we become so engrossed in our diversions that we don't notice that they are no longer making us happy anymore. Like Wile E. Coyote in *Looney Tunes*, we get our legs going so fast that it actually takes us a moment to realize that we have run right off the Happiness Cliff. Let me assure you that this never turns out well for poor Wile E.

According to the Law of Diminishing Returns, many diversions can actually be beneficial for our productivity and happiness—up to a point. Beyond that point, the diversion simply becomes a waste of time and eventually a time suck that becomes harmful to our productivity.

When I was newly married, technology played an interesting role in shaping my husband Bobo's relationship with my father. Even after I was married, my husband and dad would never have dreamed of calling each

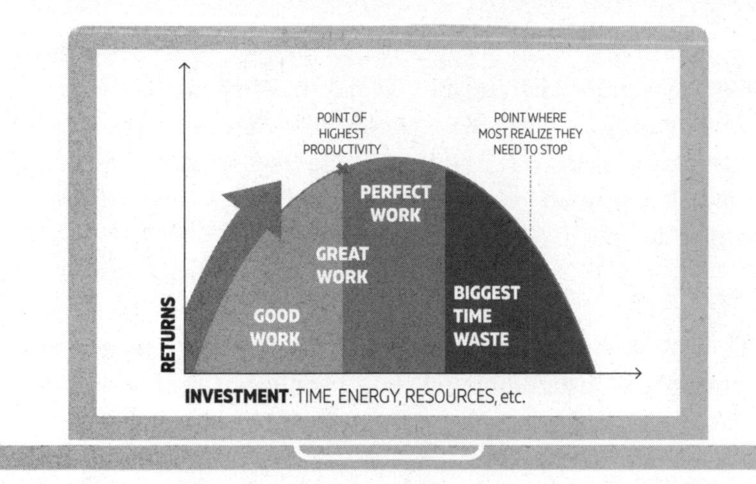

other out of the blue to get to know each other better. But one day in conversation, they discovered a mutual love of chess. So, they downloaded an app called Chess with Friends, a Wi-Fi-enabled program for engaging in friendly competition (and a little trash talking). And the next thing we knew, my mom and I kept discovering our husbands sneaking away to make their next chess move. Over the coming weeks, they began to talk multiple times a day and truly deepened their friendship. Score: +1 for tech.

At first, their newfound mutual interest was downright heartwarming. But as the weeks went on, Chess with Friends became an obsession, with both Dad and Bobo playing with random individuals all over the world to brush up on their skills to defeat the other. Yes, they got smarter and more strategic, but they also started to get a lot of flak from other relationships in their lives that were suffering. Score: −1 for tech.

At some point, though, both players began to lose interest and played less and less. However, by this point Bobo and Dad had developed a rapport and mutual understanding that facilitated a relationship in person. Score: +1 for tech.

But then, Dad discovered digital Sudoku, and Bobo started binge-watching *Breaking Bad* on his iPhone. Score: −1 for tech.

If you envision tech as a worthy opponent that you have to outsmart, you have already lost, because tech adds and it takes away; it ebbs and it flows. It also comes and goes. Like awkward introductions at a dinner party, where we barely have time to shake hands and exchange names with someone before we are introduced to the next person, the next tech comes along, leaving the former in its wake.

This push-pull relationship with technology is a zero-sum game because tech is a facilitator, not a player. We personify tech, giving devices and artificial intelligence power over choices that are ultimately ours to make, if we choose. We are the players, and as long as we believe that our behavior matters, we will win every time. When we forget that, we begin to lose focus and become less engaged.

This push-pull relationship with technology is a zero-sum game because tech is a facilitator, not a player.

PHONEY PRODUCTIVITY

Sometimes we don't use our phones authentically. We use our phones to escape our surroundings or avoid certain interactions—or for "phoney-productivity." How many of us mess with our cellphones on elevators to avoid making awkward eye contact with a stranger (and don't tell me that I am the only one who does this)? It can't possibly be because every person in the elevator suddenly has an urgent email to check simultaneously. Our social script in society says that we should always be busy or should at the very least give the impression that we are busy. And we are incredibly good at sending these signals, but all too often at the expense of our deeper relationships, which get our scattered attention because we are trying to please others. One preteen I met actually admitted that sometimes when her cellphone battery ran out, she still pretended to text away on her phone to give the illusion that she was very busy connecting with friends on social media!

In a recent survey of phone users, 54 percent of respondents said that it was fine to pick up the phone while out to dinner and 57 percent said they'd pick it up while going to the bathroom. Even worse, 33 percent of respondents thought it was okay to pick up their cellphone during sex (which gives a whole new meaning to the term "booty call").[17] But seriously, what kind of phone call could be that important to answer during physical intimacy? The alarm bells are going off right and left as we see the physical manifestations of tech addiction and the nonverbal cues with decreasing eye contact in conversation. We don't know how to teach our kids appropriate phone boundaries because we are in uncharted territory ourselves. Something is wrong with us, or it soon will be if we don't learn how to find balance in the midst of distraction.

THE STRATEGY:
GROUNDING OURSELVES WITH INTENTION

Avoiding or overcoming distraction, just like happiness, is a choice—and it's one that takes intentionality and practice to master. In the remainder of this section, I'm going to share with you some practical ways that you can ground yourself in intention.

TURN ATTENTION TO INTENTION

My six-year-old daughter, Gabri, has been working really hard to learn to read recently. One night, she picked up the book *How to Be Present* by Rana DiOrio and started reading, sounding out each and every word . . . S-L-O-W-L-Y. She spent about two minutes reading the cover, and then another two reading the title page (an exact replica of the cover). At first, I was beaming at her budding reading skills and the quality moment we were having; about ten minutes into the book, though, we were only on page two and I was struggling to sit still. I just wanted to grab the book and read it. The irony of the moment didn't escape me—I was struggling to be present long enough to read a book about how to be more present. In fact, my daughter's deliberate reading felt like nails on a chalkboard, and I had to have a serious internal conversation about why it was so hard for me to sit and listen.

Without overthinking it, I think the issue was that I had become a task-focused person rather than a process person. I strongly aligned my personal value with my productivity. And to be more productive, I needed to move quickly. To move quickly, I needed to multitask. And that book about being present was calling me to single-task, at the cost of getting my daughter to bed on time, failing to do which would prevent me from doing more multitasking—which in turn caused my foot to bounce uncontrollably each time I looked at the clock.

I didn't start out like this. As my daughter's exasperating book reading subtly suggests, children are born present and learn from adults how not to be present. Ugh. I had grown into this person, largely through a series of subconscious choices. Long ago, I jumped on the productivity treadmill and had gotten so used to it that I felt like I needed to run even when I jumped off. I have identified a few suspects in this tragedy:

- ✓ The school system, which started me on this race for education
- ✓ Society as a whole, which taught me that accomplishment was the measure of a life well lived
- ✓ Technology, which has sped up the pace of life

But the real culprit was me. My lack of being present was a choice, and it was one that I never meant to make.

The big question is: If we didn't have tech, would we really be more focused, mindful, authentic, and intentional? If not, then tech is not the fundamental problem. In fact, tech can be part of the solution . . . *but only* if we are intentional about learning how to control it.

If we didn't have tech, would we really be more focused, mindful, authentic, and intentional? If not, then tech is not the fundamental problem. In fact, tech can be part of the solution . . . but only if we are intentional about learning how to control it.

TAKE A THREE-PRONGED APPROACH

Just as my brother, Shawn, and I learned from our electricity experiment at the beginning of the last section, if you are going to plug in, you have to be grounded. Tapping into a "third prong" allows you to control the power of the technological world by focusing and channeling your energy with intention. Your third prong might be a set of beliefs, a mantra, or even just a list of rules that you live by. Regardless of the source, this grounding prong connects you more strongly to your energy source and helps you to align your efforts in a productive way. Without this prong, we are like live wires, spewing energy in all directions and putting others around us at risk in the process.

Helen Kain, co-founder of the executive-coaching firm Authentic Impact, offers a visual window into how this looks in our lives:

> *The siren call of an arriving text or email captures our attention even when we know we should be concentrating on the task at hand. We sneak a peek at dangerous or inopportune times, whilst driving, for example, or reading to our children at bedtime. Feeling vaguely guilty about this, we valiantly tell ourselves that we looked "just this once" because we are waiting for an important message. Really? Truly? Due to cognitive dissonance, our capacity for self-deception is almost as great as our love affair with distraction.*[18]

Cognitive dissonance is the mental stress we feel when we hold or act upon contradictory beliefs, ideas, or values at the same time. It's the discomfort we feel when we present one version of ourselves on Facebook, when we know that reality is actually quite different. Brené Brown, a professor at the University of Houston and author of *The Gifts of Imperfection,* describes authenticity as "the daily practice of letting go who we think we are supposed to be and embracing who we are."

The further we get from our authentic selves, the further we move from becoming our ideal selves. The way that we prevent that "mission drift" is by grounding ourselves in a set of guiding principles that inform our behaviors and habits. A few years into starting GoodThink, we found ourselves busier than we had ever imagined. In 2009 alone, Shawn had traveled to more than forty-five countries to lecture on the science of happiness, only to land in the hospital shortly thereafter with excruciating back pain. He had worked himself into the ground. Something had to give.

The further we get from our authentic selves, the further we move from becoming our ideal selves.

I take full credit as his co-founder, but more important as his bossy younger sister, for starting an important and life-changing conversation about priorities and boundaries. Fueled by his desire to share the research of positive psychology, Shawn felt driven to say yes to anyone and everyone who wanted to hear his message. I astutely pointed out that if he burned out, no one would hear his message. In fact, people looked to him as a thought leader in how to live out the principles of positive psychology—they needed him to make the hard counterculture choices. So we started to make some very tiny changes in the way that we did things.

One of the simplest changes was to write an authentic "away" message when he went out of town. Instead of the innocuous note alerting people that he would be "out of the office until Monday," he began to expound on what he was doing. "I'm taking a vacation with my family and trying to be more present by stepping away from technology for a week." At first, he feared that partners would be annoyed that he was unreachable, but the reaction was exactly the opposite. Shawn got a flood of emails (which he read after vacation) thanking him for practicing what he preached, for validating what others believed we all should be doing, for giving others the courage to try the same thing themselves.

Shawn made a trade-off that aligned with his values, favoring authenticity and quality time over productivity, and the payoff was exactly what he desired—closer connection with the organizations that shared his vision for work-life balance. Not all trade-offs are this straightforward and simple, but when choices are grounded in intention and thoughtfulness, you can be assured that the choice is one that you will not regret.

Questions to consider:

- ✓ What values, principles, or beliefs comprise your third prong?
- ✓ Is your third prong plugged in firmly, or are you letting other distractors zap your energy and derail your efforts?
- ✓ Are there practices like journaling or meditation that could help you realign your day-to-day choices with your priorities?

SET YOUR INTENTION

In 2014, I was invited to spend the afternoon with Oprah as GoodThink filmed a two-part series titled "The Secrets of Happy People" for the *Super Soul Sunday* show. At the end of each show, Oprah asks all her guests a series of rapid-fire questions, including questions like, "What is hope? What is belief? What is the soul?" My favorite question of all that she asks on her show, though, is, "What is your intention?" When you start out your day, why are you alive and what do you hope to accomplish? Without setting our intention, we quickly get swept into the onslaught of tasks, checklists, and priorities that other people have set for us.

When you start out your day, why are you alive and what do you hope to accomplish? Without setting our intention, we quickly get swept into the onslaught of tasks, checklists, and priorities that other people have set for us.

I believe there are four ways to harness the power of intention: (1) actively choose your own adventure based on your values and personality, (2) understand others' intentions, (3) focus on tuning in rather than zoning out, and (4) bring your priorities into the foreground of your life. Let's look at each one.

Actively Choose Your Own Adventure: How You Interface with Technology

Imagine a child who grows up in a small town. He spends his days exploring and going to school and meeting people. Every day, his understanding of the world grows—in other words, his brain is completely malleable. When he grows up, though, he decides it's time to move to the big city. So he sets off on the dirt road to his future home. However, along the way, he encounters a strange shiny object in his path. Is this object part of an ambush? A normal part of the scenery? Or a fortuitous opportunity?

Each of us is on our own adventure, and we have all encountered shiny objects along the way. When faced with an unknown element in our path, many of us unconsciously respond with an emotional or gut reaction. But by delving into this book and raising your awareness about technology, you now have the opportunity to *consciously* choose how you want to engage in the future. How much do you want technology to be a part of your life? What are your boundaries and why? How do you plan to sift through which technologies might be helpful or harmful? These are the personal choices ahead, and every path leads somewhere interesting. The only mistake you can make along the way is not to ask these questions at all.

In an interview with *Business Inside,* twenty-five-year-old Ben Brast-McKie explained why he finally decided to ditch his cellphone and hasn't looked back for four years:

> *What changed? First I noticed that I had developed many compulsive tendencies. I would feel the "itch"... I would also feel the "leash" if you know what I mean. The phone became a constant interruption. It doesn't matter that the people interrupting are people I knew and loved. It was still an interruption. I noticed myself being interrupted, breaking the natural cadence of my conversations. I noticed others doing the same to me. At first this was socially "rude." People would apologize but do it anyways. But this didn't last very long. Now it is accepted and expected.*

My good friends who came to visit me one weekend echoed these sentiments. Ron and Angeline are both surgeons in California and have two children. Given their busy work schedules, Ron and Angeline were excited just to have a lazy weekend at our Dallas home. After chasing kids around the house for thirty minutes, I finally plopped down on the couch to catch up on life with Angeline. Despite five children running around the house like banshees, Angeline exuded calm. I, on the other hand, exuded a fuzzy caterpillar that had been rubbed down with a static-electric balloon. Secretly, I had been observing Angeline all weekend because she was one of the few people I know who still used a relic of the '90s: a flip phone. Despite living one block from the new Apple Rocketship campus in the heart of Silicon Valley, Angeline was a self-described "conscientious objector" to smartphones.

Trying not to sound judgey-judgey (a symptom of my jealousy that she was able to sit down and read more than a page of a book at a time), I casually asked Angeline why she chose not to get a smartphone. *How did she survive?* Angeline's serene face curled into a smirk, as she admitted that her colleagues hated that they couldn't always reach her but that smartphones just held no allure for her. "They don't have anything that I need," she explained. "And I don't want my kids to constantly ask for my device as their personal form of entertainment—I want my children to read books and not play games. Sometimes I worry that I'm creating a child who may not be socially adjusted if they are not attached to a phone, but so far they seem just fine."

Inspired by her attitude, I decided to experiment a bit with her approach. When I took my six-year-old Gabri to the grocery that afternoon, she immediately asked for my cellphone while she sat in the grocery cart. It certainly would have been easier and faster to shop with her self-entertained, but instead I declined giving her my phone and handed her a book instead. After a bit of grumbling, she was soon engrossed in her book and I sped through my shopping trip. As I began checking out at the counter, the clerk spontaneously complimented my daughter on reading instead of being on a device. She explained that she saw parents come through her lane all day, every day, with their children glued to a device (I guiltily grinned). "When I was a child, we played outside for hours," she said. "Now my children are attached to the electric socket in the wall, only wandering as far as their charger will go. If I tell the kids to go outside and play, they only leave for long enough to let their devices charge." This sentiment is one that has been echoed over and over as I interviewed people for this book and conducted my research. What social scripts are we unintentionally setting up for the future by not consciously thinking about the eventual implications of our actions?

In my research, I found that most individuals tend to fall into one of three personas: Embracers, Accepters, or Resisters.

Personas	Characteristics
Embracers	Love to be on the cutting edge either for information gathering or status-seeking

Accepters	Have a nominal interest in new tech but wait for products to hit mainstream
Resisters	Lack desire, motivation, or financial resources to try new technologies

Despite the lesson learned from the grocery-shopping experience with my daughter, as an Embracer I must admit it was incredibly tempting to write this section as an entreaty to all Resisters to try a Day in the Life of an Embracer (in the back of my head, I hear the singsong words of Dr. Seuss in his beloved children's book *Green Eggs and Ham*: "You do not like them. SO you say. Try them! Try them! And you may."). But I resisted because the more objective side of me knows that one persona is not better than another.

All three personas have advantages and disadvantages. Embracers love to be at the forefront of new trends, but their fascination can become an expensive hobby. Accepters like to wait until new technologies become cheaper and have fewer bugs, but they find themselves constantly following trends rather than creating them. Meanwhile, Resisters may save time and money on technology but often lag behind societal trends. Nevertheless, all three personas can become incredibly distracted in the process of trying to use or even trying to resist technologies (my friends who insist upon using a flip phone instead of a smartphone spend infinitely more time crafting each text message using the limited alphanumeric keyboards). Likewise, all three personas can become more grounded (even my Embracer friends who use Headspace daily to develop their meditation skills). Moreover, sometimes these personas are domain-specific. For instance, my friends Ron and Angeline might be Resisters to tech in their home but are full Embracers when it comes to the operating room. My parents, on the other hand, are Accepters of new technology at work but Resisters when it comes to online banking.

Behind each list of characteristics is a set of values that each persona holds dear, a unique perspective that shapes their choices and actions as consumers. Whether driven by faith or a code of ethics, values ground us both consciously and unconsciously. By recognizing which persona you are and why, you begin to set your intentions moving forward, in a process that fuels your attitude and approach to building key skill sets to improve your happiness and well-being in the future. Consider: Why might you have adopted your persona? Do you find that you embrace technology in some areas but reject it in others? Understanding your own motivations helps to solidify your perspective and clarify choices for purchasing and interacting with tech in the future.

Understand Others' Intentions

As important as it is to understand your intentions using tech, it is equally important to understand the intentions of data-collecting companies as well, lest you find yourself supremely unhappy as the victim of sensitive information leaks. In the era of big data, where privacy law is being written as we speak, we have to be vigilant about who is capturing information about us, where it is going, and how it is being used. I advocate using the three Ps to bolster your personal security: privacy, people-finding, and passwords.

Privacy—Check your privacy settings on social media to ensure that you are limiting your sharing of information to people you really know and trust. And be that weirdo who reads every privacy statement before signing up for a new website or service. If the privacy statement is hard to find or does not exist, don't sign up!

People-finding—Did you know that dozens of search engines like whitepages.com, spokeo.com, and peoplefinder.com serve as data brokers, selling access to a full profile about you, including your current and past addresses, phone numbers, and even a list of your closest family members? Most of these services offer a way to opt out of being listed, which may be well worth your time.

Passwords—Part of well-being in a digital world is having the peace of mind to know that your passwords are safe, your data is protected, and your online profiles are secure. If you have concerns, consider using 1Password.com or iPassword.com, both of which are designed to safely handle your most sensitive online data.

Though it's good to be vigilant about your data, not all data-gathering sites are malicious. Storm alerts, pet-finders, medical-emergency apps, and health trackers have all contributed to a new level of safety and security that I would never want to give up now that I know it is possible. In the future, these apps will only become better and more synchronized to give important information to you in a more coordinated fashion.

To understand how data can be used for good in the future, I caught up with Chronos app founders Charlie Kubal and Dylan Keil. The Chronos app was designed as a "passive time tracker," recording details about how much time you spend on your phone, which apps you use, which online locations you frequent most often, and more. The goal was to give the users insight into their own behavior so they could be more intentional about using their time well. Chronos was recently acquired by Life360 and will be part of a social tracking app to stay in touch with your family and close friends. Charlie and Dylan explained that they invested a lot of time and thought in developing their privacy policies so that users would feel comfortable and safe with the way their data was being used—ensuring that users always own their own data; that they always have the ability to opt in; and that there is transparency in how their data is being used.

USE TECH TO TUNE IN, NOT ZONE OUT

It's easy to get frustrated by the phone incessantly ringing, pop-up messages with new emails, and dings alerting you to new text messages. Tech helps us communicate faster, but it's also become an albatross around our neck. Researchers are just beginning to delve into the topic of how tech affects happiness and emotional development in the long run. One study at Stanford University explored the online habits of girls ages eight to twelve by having them surf the internet for up to five hours and then self-report about their happiness levels and social comfortability.[19] The study found that the more time the girls spent on screens, the more they described themselves in ways that suggest they are less happy and less socially comfortable than peers who say they spend less time on screens.

Another study found that the mere presence of a cellphone during a face-to-face conversation reduces feelings of closeness, trust, and relationship quality, even if the phone is not being used.[20] In her *New York Times* bestselling book *Alone Together*, Dr. Sherry Turkle, founder of the MIT Initiative of Technology and the Self, expresses concern that we are setting ourselves up for trouble—we are so used to communicating through devices that we are losing the ability to connect on a deep, personal level in real time.

While these findings *should* give us pause, there is another story that can be told here as well, namely one that focuses on how technology can (and is) being used to *improve* communication. Think of the numerous geographically divided families that can now communicate on Skype for a fraction of the cost, of deployed soldiers who can read bedtime stories to their children over FaceTime, of children of prisoners who can now communicate daily with their parents rather than waiting weeks for a single letter or a short phone call. These modes of communication are not replacements for other forms of traditional communication; rather, they are entirely new modes of communication that provide additional inroads for dialogue and relationship building.

While sensationalist stories abound about how technology is eroding the very fabric of our society and how screen time has become the digital equivalent of heroin, I prefer to take a more measured approach, continually going back to the science to understand emerging trends. Interestingly, I learned that one of the most widely quoted studies on communication and technology has since been replicated and updated to reveal fascinating new insights. The original study, performed in 1998 by Carnegie Mellon's Robert E. Kraut, tracked the internet use of volunteer families with high school students. Kraut found that the more they used the internet, the more their depression increased, and the more social support and other measures of psychological well-being declined.[21] However, in 2002 Kraut decided to repeat the same experiment, this time paying careful attention to whether students were interacting with individuals with whom they have strong ties (close friends, family, etc.) or weak ties (strangers, acquaintances). By looking at the nuance of *how* time was spent on the internet, he found that the students who interacted with strong ties showed a *decrease* in depression, a reduction in loneliness, and an increase in the level of perceived social support.[22]

Another study of more than 600 individuals on the internet found that "50 percent of these participants had moved an internet relationship to the "real-life" or face-to-face realm. Many of these online relationships had become quite close—22 percent of respondents reported that they had either married, become engaged to, or were living with someone they initially met on the internet. In addition, a two-year follow-up of these respondents showed that these close relationships were just as stable over time as were traditional relationships."[23,24]

Keith Hampton, an associate professor of communication and public policy communication at Rutgers University, argues that the idea that we interact either online or offline is a false dichotomy. Through his studies, he has become convinced that social media and the internet are actually drawing us closer together—both online and offline. "I don't think it's people moving online, I think it's people adding the digital mode of communication to already existing relationships," he says. The more different kinds of media that people use to interact—phone, email, in-person, text, Facebook—the stronger their relationships tend to be. Similarly, a 2012 Pew research study of more than 2,200 individuals in the US found that 55 percent of internet users say their email exchanges have improved their connections to family members, and 66 percent say the same thing for significant friends.[25] Sixty percent of users cite email communication as a primary reason for this improvement.

Fostering stronger human communication is the reason Peter Steppe, founder at OEX Inc, created Campfire, an app designed to trigger greater human connection using phones. The app simulates a campfire experience among friends (where phones on the table = the campfire). One individual, called the "firestarter," initiates the first spark of flame and draws other friends in to a space and place to relax and check in through face-to-face communication. The longer that your phones remain on the table, the stronger the flame and the more points earned. Peter explains that the app "feels a bit like going outdoors, yet at the urban table."

When Peter first started the app, he intended the tool to be a means of getting young people to put their phones down (what he terms a "calm tech state"). However, he quickly realized that that approach was shortsighted and not attuned to the larger issue—finding ways to help people of all ages connect to each other in a smartphone era dominated by so many "alone together" situations. Rather than replacing or avoiding tech, Peter decided instead to create a dynamic app that facilitates and encourages "in the moment" interactions with friends and family through thirty-minute daily-dose situations within the structure of daily life. Peter believes that strong ties lead to the most sustained feelings of happiness, but that weak ties are also valuable because they are the seeds of new friendships. Rather than cutting out weak ties, he advocates that we think about our ties with a connection-growth score, a challenge of sorts. How much can you increase the depth of your connections?

BRING YOUR PRIORITIES TO THE FOREGROUND

Did you know that 1+1=3? This might sound like fuzzy math, but it's a concept I can prove to you. If you hold your index fingers up together, you clearly see two fingers. However, if you move one finger closer to your nose, focusing on the finger moving closer to you, you begin to see three fingers. This fascinating dynamic, called stereopsis, occurs because as your eyes try to fixate on the finger that is closest to you, everything in the background becomes blurred. Your perspective shifts from spatial (seeing objects on a horizontal or flat plane) to temporal (seeing objects relative to depth).

This is the same technology that enables 3D glasses to create depth of field by merging images, and will soon enable screens to create the same effect without specialized glasses using advanced stereopsis. In fact, a company called Alioscopy won the French Worldwide Innovation 2030 competition by debuting its glasses-free 3D screen, meaning that, in the near future, billboards and computer screens will be even more distracting, with images literally jumping out at us.[26] Thank you, technology. Shows like *The Walking Dead* already freak me out. Soon scary shows will be even scarier.

In 2015, I had the opportunity to go to Milan to visit the World Expo, which featured a theme of "Feeding the Planet, Energy for Life!" My favorite exhibit at the Expo was centered around the future of food and included a supermarket of the future, a mechanical bartender, and even a futuristic model kitchen, complete with a refrigerator that told me which foods I should eat based on my biometrics and a fingerprint scan. I also tested out virtual reality Oculus goggles for the first time, weaving my way through a program that guided me through an immersive visual Wikipedia of nutrition information for every edible object in my field of vision. The novelty of the idea was so fun and intriguing, but the deeper I delved into this world, the less aware I was of the external, real world. My information awareness came at the cost of my self-awareness, leading me to be totally comfortable flailing my arms around in public with giant goggle headgear on. My self-consciousness disappeared because my consciousness of the world around me disappeared. While immersive experiences like this one can be fun for a brief time, ultimately I want not just to maintain my consciousness, but actually to use technology to *increase* my awareness and understanding of the world. To do so, I need to stay grounded in the face of distraction; the feat will require my setting some healthy boundaries for my interaction with technology in the future.

SET HEALTHY BOUNDARIES

As technology floods our lives, literally pouring in through every crevice in our homes, it becomes increasingly difficult to keep our focus on other things. We vacillate between being constantly distracted by technology and wanting to be completely disconnected from it. But there is a third option, and that is to learn to set better boundaries in our lives. There's an old saying, "Good fences make good neighbors." Having good boundaries on your use of technology will make you a better family member and co-worker (and keep you from flailing in public).

Happy Hacks to Get You Started

1. **Turn off notifications.** As Tony Schwartz and Jean Gomes write in *The Way We're Working Isn't Working*, "We each have one reservoir of will and discipline, and it is depleted by any act of conscious self-regulation, whether that's resisting a cookie, solving a puzzle, or doing anything else that requires effort." Unless you have superhuman powers to resist alerts that have been carefully designed to grab your attention at all costs, do yourself a favor and turn off as many notifications as you can. In a recent study, researchers asked individuals to keep their phone notifications on and within reach for one week; the

following week, those same individuals were asked to turn off notifications and keep their phones out of sight.[27] The study found that individuals who kept their notifications on reported significantly higher levels of inattention and hyperactivity, which in turn predicted lower productivity and psychological well-being.

2. **Limit your information feeds.** To the greatest extent possible, limit your checking of information feeds (email, social media, news, sports) to three times a day. A recent study found that checking email less frequently significantly decreased stress, which then paid off by increasing a sense of meaning, social-connectedness, and even sleep quality.[28]
3. **Protect your brain's consolidation time.** Our brains use downtime to download and consolidate all of the information that they receive during the day. If we fill all of our downtime with digital distractions (surfing Facebook, posting on Instagram, playing games on our phones, or even reading e-books), the brain has no time left with which to process the world, chunk information, and form long-term memories. As digital strategist Tom Gibson writes, "We need to understand that 'on' is impossible without 'off,' and that the distance between the two needs to be made closer; like the beats of a heart or the steps of a runner."[29] Instead, try to establish device-free brain breaks to help your brain recharge (right before bed/after waking up, during a walk, or even during play) and refocus. The National Sleep Foundation and Mayo Clinic recommend that you abstain from using digital material one hour before bed specifically to block the release of neurotransmitters that energize your brain and keep you from entering a restful state that your body needs.
4. **Set up safeguards.** For parents, privacy and safety are huge issues in the Digital Era. Until recently, monitoring children's use of the internet was cumbersome and overwhelming. However, recent developments like the

KidsWifi router have made this process infinitely easier.[30] In just two minutes, you can set up KidsWifi to filter, monitor, and control all of your kids' online devices and even those brought over by friends. You can even establish Wi-Fi-free times in your household (during dinner or at different bedtimes for different children).

5. **Model digital citizenship.** Establish some personal standards for hard-and-fast rules for your use of technology when interacting with others: Look up from your computer when someone walks into the room, take out your ear buds to say hello, and close your laptop when having a conversation.

The National Sleep Foundation and Mayo Clinic recommend that you abstain from using digital material one hour before bed specifically to block the release of neurotransmitters that energize your brain and keep you from entering a restful state that your body needs.

POST YOUR GOALS

In addition to establishing and maintaining boundaries, you can set up visual reminders of your priorities. In *Before Happiness*, Shawn Achor explains:

> *Your mind is a goal-oriented machine, subconsciously making assessments about how far away a goal is (proximity), the likelihood of achieving it (target size) and the effort (thrust) required . . . but these variables are based largely on our perception. Unless you can see into the future, you can't possibly know how far away your goal is, your likelihood of attaining it or how much effort it will take, but you can control how you perceive the proximity of the goal and the effort required to succeed.*

The best way to bring your priorities to the foreground is by visually posting them somewhere that you spend a significant amount of time. A few years ago, I got inspired to write down my goals in a really visible place in my house, so I picked a wall by my kitchen and painted it with black chalkboard paint. I then proceeded to write down a list of my summer goals, so that every time I walked by, I would be reminded of my intentions. The added benefit was that my family decided to make goals as well, and we started to hold each other accountable. "Did you learn how to do a handstand yet, Mommy?" my oldest daughter would ask me every day. "No," I answered snarkily. "Did you teach me how yet?" Touché.

Even visitors who came to our house started to get inspired, making their own lists, too! In our world of competing notifications and reminders, going "old school" and manually writing your goals on a vision board, a wall, a mirror, or even Post-it notes can significantly increase the likelihood that you will follow through on them. Bringing your priorities to the foreground eliminates distractions and helps you to stay grounded to become the person that you want to be.

SUMMARY

Although our attention spans might be shorter than those of goldfish, we can learn to become less distracted and more present in our lives, so that we can tap into a greater sense of flow, fully immersed and engaged in whatever activity we might be doing.[31] We must tap into our third prong to ground our choices about when, where, why, and how we engage with technology so that we can better channel our energy toward creating a happier future.

Stay grounded in the midst of change by:

- ✓ Utilizing the "third prong" (your guiding principles and values) to focus your energies
- ✓ Reducing distractions to increase productivity
- ✓ Actively choosing how you want to respond to technology: resist, accept, or embrace it

✓ Understanding others' intentions as well as your own
✓ Focusing on tuning in, not zoning out
✓ Bringing your priorities to the foreground
✓ Posting your goals somewhere visible

STRATEGY #2
KNOW THYSELF

HOW QUANTIFYING YOURSELF HELPS ELIMINATE LIMITING BELIEFS ON YOUR POTENTIAL

In ancient Greece, philosophers believed so strongly that self-knowledge was the key to human potential that they inscribed the phrase "Know Thyself" onto the sacred Temple of Apollo. Since that time, philosophers, religious leaders, and authors alike have mused about the nature of humanity and our sense of self. What are humans made of? How do humans experience pain? What are emotions and why do we have them? Until recently, most of the conclusions that were drawn came from external observation or speculation. However, thanks to technology, we now have the ability to connect our external and internal worlds in ways that Socrates or Plato couldn't have imagined.

We have evolved from using kitschy mood rings to reveal our emotional state to having real tech to understand what's going on inside our bodies on an intellectual, emotional, and even molecular level.[1] We are witnessing a new era in which people can actually get a personal, real-time snapshot inside their bodies—organs, cells, DNA, and a whole molecular universe of other tiny structures. With MRIs, we have literally been able to peer into the brain to see how stimuli like stress affect decision-making, which then triggers other physiological responses like increased heart rate, sweat, and headaches.

> We have evolved from using kitschy mood rings to reveal our emotional state to having real tech to understand what's going on inside our bodies on an intellectual, emotional, and even molecular level.

Just as science has helped us understand some of the implications of negative stimuli, so it is helping us understand the effects of positive stimuli as well. For instance, today's athletes are now bigger, stronger, and faster than ever before because they have learned to hone their bodies and minds to reach more of their potential. What if we adopted that same rigor and discipline to developing our minds, from a neuron level to an entire system level? Self-knowledge is power. In this second strategy, I will show you how technology can help you understand your full potential by helping you to track key information about your habits and then use that information to fuel your growth.

THE CHALLENGE: RECOGNIZING LIMITING BELIEFS

Despite being highly sophisticated human beings, all of us make less than optimal decisions from time to time—decisions that hold us back from reaching our full potential. The reason is that we often lack basic knowledge of ourselves for our brain to properly model out whether an action will be good or bad for us.

The problem is that sometimes we have just enough information at our fingertips to think that we have thought through an idea well; however, on closer examination, there are major gaps in our thought process. This phenomenon is known as illusory contours and is based on heuristics—cognitive processes that help to make quick sense of the world, even though they may prove to be faulty. To show you how this works, take a look at the images below. As you look at each picture, your brain begins to add shapes and lines that are hinted at but don't actually exist.

Illusory contours by Peter Hermes Furian/ShutterStock.com.[2]

The brain does the same thing each time we are faced with a new challenge. The brain logs key information about the task and then fills in the gaps with illusory knowledge—that may or may not be accurate—to draw conclusions. It turns out that, most of the time, we are flat-out wrong. In fact, roughly 50 percent to 80 percent wrong.[3] You might be thinking this very thing about your spouse, and now you have scientific proof! For this reason, psychologists and authors like Dan Ariely say that we are actually "predictably irrational" in our decision-making. If you need a concrete example of how you think irrationally, feel free to ask a parent, sibling, friend, or perhaps better yet, an ex-boyfriend or ex-girlfriend. I'm sure that any of them would be happy to provide you with a detailed list of examples. (Don't ask your significant other; better to let him or her continue believing that you think you are always rational.)

If we truly want to learn to overcome challenges and strive after our potential, we have to learn to recognize the illusory knowledge in our environment that causes our limiting beliefs. Only then can we begin to reframe our thought processes so we can mindfully begin to fill in the gaps where we might need more facts and information so that we can make empowered choices. To demonstrate how illusory knowledge plays out in our lives, let me tell you a story.

Two years ago, a group of girlfriends invited me to walk a half marathon with them in the Outer Banks. Now, mind you, I have always *hated* running, but I thought walking wouldn't be too bad. Plus, I was really craving a girls' weekend away, so I signed up on a whim. About one week into training, my "friends" decided we should *run* the race instead. What?! I panicked. I had never run longer than one mile before. In elementary school when I was five years old, they asked me to run a mile. About a quarter of the way in, I almost collapsed because I couldn't breathe. Since then, I had resisted running at all costs.

But, as I now understand, I had a limiting belief based upon a single data point: my loss of breath at age five. I didn't know whether that was an asthma attack, or if I had been born with one lung (a condition somehow undetected by every doctor in my life), or if I was just perhaps not in the practice of running. So I set out to log and quantify a renewed attempt at running, using the MapMyRun App, which geo-tagged my location as I ran my first five minutes. During my first training session, I noted that my breathing began to get difficult at exactly .39 of a mile (at the four-minute mark). I also logged how long it took me to get my breath back (six minutes), how my legs felt, what my heart rate was, and in what kind of weather conditions I was running. Two days later I tried again. I made it the same amount of time before my breathing became labored. I logged that information. Then over the next few runs, my run duration began to increase, my speed got faster, and my breathing recovered sooner. I suddenly had real knowledge: I could run at least four minutes without breathing hard—but my full potential was still unknown.

Six weeks later (the halfway point in my training), picture me running five miles in the mountains *on vacation*. Yup, that was me. This transformation did not happen overnight but was rather a series of little battles and choices along the way that required me to rethink what I thought I knew about myself. By the end of my training, I found myself sprinting to the finish line of the half marathon, having run the whole 13.1 miles without stopping.

This feat continues to be one of the proudest moments in my life, not just because I finished the race, but even more so because I overcame limiting beliefs that I had struggled with for years. Quantifying my behavioral patterns and having more than one or two data points changed everything.

Illusory knowledge threatens to derail our decision-making by skewing our perception of reality. These traitorous ideas are often hidden in the smallest, quietest thoughts in our heads, whispering falsehoods, spreading seeds of doubt, and holding us back from achieving our full potential. The key to making better decisions is taking the time to look thoughtfully at the details that shape our larger environment. You've probably heard the expression, "Sometimes it's hard to see the forest for the trees." It's worth stopping to realize that there would be no forest without trees.

Illusory knowledge threatens to derail our decision-making by skewing our perception of reality.

At GoodThink, we define optimism as the belief that our behavior matters. And when we latch on to this idea, we begin to take ownership of these small moments, recognizing that they are not just fleeting thoughts but critical choices that shape our future. Until you believe that your behavior matters, change is virtually impossible. Every day, you have the opportunity to make an active choice that your behavior *does* matter, both for your success and happiness—not in some distant future, but right now, right here in your life.

At GoodThink, we define optimism as the belief that our behavior matters. And when we latch on to this idea, we begin to take ownership of these small moments, recognizing that they are not just fleeting thoughts but critical choices that shape our future. Until you believe that your behavior matters, change is virtually impossible.

THE STRATEGY:

MAGNIFYING YOUR MICRODECISIONS

Over the years, my family has made endless fun of my love of Excel spreadsheets. "Hey, Amy, we are heading to the movies. Want to make a spreadsheet about what we should see?" To be fair, I might have earned this reputation by using spreadsheets to evaluate just about every major turning point in my life, including whether or not to get married, to make a career change, to move, or even to have a third child. Yet I justify my countless hours perseverating over options because "it's a big decision!"

However, when it comes to the little decisions in my life, I tend to shoot from the hip. As a working mother of three kids and two dogs, I often start my day on autopilot, which means that sometimes my decision-making sets out on a trajectory that may not be the best use of my time or energy. I unconsciously make small decisions about whether to procrastinate, whether to skip exercise yet again, whether to return that phone call, whether to let my children watch another show so I can get things done. And I do this without regard to the compounding effects of these small decisions in my life. I call these inflection points "microdecisions," and they are the building blocks of habits. Microdecisions will either be the start of your slippery slope down a mountain or the pick that gives you a stable foothold to keep climbing up the mountain.

Microdecisions will either be the start of your slippery slope down a mountain or the pick that gives you a stable foothold to keep climbing up the mountain.

These small decisions, which feel disjointed and innocuous, are the biggest determinants of our productivity, and ultimately of our happiness as well. Les McKeown, CEO of Predictable Success, goes so far as to call microdecisions "the most overlooked key to leading a successful company." In his blog for Inc.com, he explains how we often judge companies by the small decisions that front-line employees make: whether they smile, look you in the eye, pay attention to detail, show up for work on time. The average employee makes between 25 and 200 microdecisions every day. McKeown crunches the numbers: "For a business employing just 10 employees, that's a cloud of between 250 and 2,000 succeed-or-fail events that is generated every single day. Get most of them right, you win. Get most of them wrong, and you lose."[4]

Often microdecisions feel divorced from larger trends because we convince ourselves that they are so small in the grand scheme of things that they do not make a difference. Yet the biggest decisions that we face in life are almost never isolated moments; rather, they are shaped by a series of microdecisions that we have made over time, whether consciously or unconsciously. Malcolm Gladwell calls this cumulative effect the "tipping point," or that magic moment when an idea, trend, or social behavior crosses a threshold, tips, and spreads like wildfire.[5] Just like the Butterfly Effect (where it is said that a butterfly wing beating on one side of the world will create change on the opposite side of the world), these microdecisions have a cumulative effect that sets a path in motion and dramatically shapes the landscape of our future.[6]

> The biggest decisions that we face in life are almost never isolated moments; rather, they are shaped by a series of microdecisions that we have made over time, whether consciously or unconsciously.

Your microdecisions have direct impact not only upon your success, but also upon the culture at your organization. If you are among the 87 percent of employees worldwide who feel disengaged from their jobs, you can at the very least take comfort that you are not alone.[7] But take a moment to imagine what would happen if 87 percent of the world began to see the workday as a series of microdecisions, or discrete opportunities to invite positive personal change into their own lives as well as the culture and environment at work.

How are your microdecisions shaping your workplace, community, or family? Are your choices driven by fear or facts? Do you complain about your environment, or do you set out with an intention to fix it? In this strategy, I'll describe four ways in which you can use technology to help you to magnify your microdecisions to achieve more of your potential.

CAN YOU MEASURE UP?

The ancient philosopher Protagoras famously declared that "man is the measure of all things," but perhaps a more fitting statement for our times is "the measure of all things is man." Fascinated by this idea, thousands of individuals in more than thirty countries have now joined an online movement called the Quantified Self.[8] As part of this movement, individuals commit to "lifelog" or track their own personal data and then share their results with the world in an effort to better understand human nature.[9]

> The ancient philosopher Protagoras famously quipped that "man is the measure of all things," but perhaps a more fitting statement for our times is "the measure of all things is man."

But who, you wonder, would lifelog rather than just live in the moment? You might be surprised to learn the answer: you! Chances are that, at some point in your life, you have already lifelogged in some form or fashion, whether with a sports watch, a pedometer, a fitness tracker, or even an iPhone—it's tracking steps now! According to a Pew internet study, 69 percent of Americans are already tracking at least one health metric online, a number that is quickly rising.[10] Although lifelogging is most commonly used to track personal health, it can also be used to measure air quality (Koto Air), energy use (Nest), and even memories (Snapchat Spectacles). Lifelogging, in short, is just a method of studying yourself and the world around you over time.

Long before we had smartphones and computers, some of the world's greatest minds had caught on to the idea of lifelogging. Leonardo da Vinci was known always to carry two notebooks to capture observations about himself and the world around him. Similarly, Benjamin Franklin kept a daily journal to track his progress in achieving thirteen personal virtues that would lead to moral perfection. In his autobiography, Franklin wrote, "I was surprised to find myself so much fuller of faults than I had imagined, but I had the satisfaction of seeing them diminish." I'm not sure I want to know how many moral faults I have, nor would I want to publish them, but fortunately, there are other ways to go about lifelogging for self-improvement.

Today we are fortunate enough to be living through a digital renaissance, literally a rebirth in the way that we understand and study ourselves.

Today we are fortunate enough to be living through a digital renaissance, literally a rebirth in the way that we understand and study ourselves. Following in the footsteps of da Vinci and Franklin, a number of individuals are leading the way for the rest of us to see the benefits of lifelogging and lifehacking (cracking the code of what makes us tick better and more efficiently). Prominent among lifeloggers are Chris Dancy, who uses a number of monitoring devices to record data about himself;[11] Nicholas Felton, who has transformed his lifelogs into complex (and informative) data graphs;[12] and marketing genius Tim Ferriss, who fused the ever-growing interest in productivity with gamification to produce a *New York Times* bestselling book and program called *The 4-Hour Workweek*. Far ahead of their time, these individuals have embraced a counterculture lifestyle to raise their self-awareness and to hone their lifestyles.

Lifelogging doesn't have to been time-intensive or complicated; it is simply a means of tracking your habits online. The bonus of doing it is that you receive automated insights into your data. What gets me excited about lifelogging is the prospect of being able to create a renaissance in my own life, using the small insights to create positive changes in my life.

As Victor Lee, an assistant professor at Utah State University, writes, "When you look at data about yourself, you don't strictly see points or dots or bars. You see a depiction of an experience or activity that is already intimately familiar . . . If it's exercise data, you remember what certain moments felt like. You have some expertise on how your body works and that creates some expectations and support for thinking with data." Context helps fill in the gaps, illuminating patterns that can be hard to recognize or can be clouded by emotions.

What gets me excited about lifelogging is the prospect of being able to create a renaissance in my own life, using the small insights to create positive changes in my life.

Lee later went on to bring this method of continuous improvement into high school classrooms to improve data literacy and expose students to "more authentic forms of inquiry."[13] Of course, data and assessments are not new to the classroom environment, but Lee believes that all these tests and benchmarks are missing the point. He explains: "If we focus data-driven efforts solely on assessment, without developing student-empowering technologies that give learners insights into their progress, then we are failing the real goal as well as our students." Instead, he advocates teaching students the language for data analytics (outlier analysis, data visualization, and pattern recognition) and the process for learning (how and why things happen).

Research is useless unless it is lived.

At GoodThink, we often say that research is useless unless it is lived, and Steven Keating is the perfect example of how lifelogging can be literally lifesaving. As a college student, Steven joined a research study in 2007 that included an MRI scan. Curious about his own results, he asked if he could retain a copy of the raw data.[14] The study revealed that Keating's brain had a slight abnormality, right near his brain's smell center. He was advised to get reevaluated in a few years. So in 2010, Steven returned for another MRI, which showed no changes, suggesting that the abnormality was most likely benign. However, in 2014, Keating noticed that he was experiencing a phantom vinegar smell for about thirty seconds every day. He requested a third MRI, only to discover that the abnormality had grown into a tumor the size of a baseball. Fortunately, Keating was able to get the tumor surgically removed in under a month and was back on campus less than a week afterward. To aid in his self-study, Keating had his entire ten-hour surgery videotaped so that he could learn from the experience. By actively studying his own cancer as a scientific problem, Steven wound up saving his own life. As a graduate student now in the MIT Department of Mechanical Engineering, Steven is using his approach toward self-study and continuous learning to advance knowledge for others as well.

Like Steven, we have the fortune and capacity now to study ourselves on multiple levels using wearables that can help track information related to our health, our productivity, our well-being, and our overall happiness. Lifelogging helps us to study and connect the dots about how seemingly unrelated factors like weather, air quality, and even time of the day might be affecting our mood and productivity. These insights heighten our awareness, enabling us to shift away from unconscious negative behaviors toward more intentional positive choices, which ultimately propels us toward greater success and happiness.

By tapping into our innate curiosity and desire to problem-solve, lifelogging can provide valuable insight to increase our potential and future happiness. According to a study by Rackspace in 2014, employees wearing wearables at work became 8.5 percent more productive and 3.5 percent more satisfied with their jobs.[15] This is amazing because, of course, wearables do not directly change behavior; they simply raise awareness and provide insight. Yet, they provide a clear path for employees to improve personally and the benefits cascade into their professional lives as well.

By tapping into our innate curiosity and desire to problem-solve, lifelogging can provide valuable insight to increase our potential and future happiness.

Assessing Your Progress

How do you know if you are actually getting happier over time? If your increased well-being is actually leading to greater productivity or success? While many apps have built-in metrics to help you benchmark and track your progress, here are some guiding questions to help you assess your progress:

1. What changes have you seen over the last week? Month? Year?
2. What can you learn from these changes?
3. Did you notice any unusual results or trends?
4. What additional information would be helpful to know?
5. What is one tweak that you could make to improve your results?

ACTIVATE YOUR INFORMATION FOR TRANSFORMATION

Information is one of our greatest sources of motivation and power. However, as my brother Shawn often says in his talks, "Information alone does not cause transformation." Data is only useful and relevant when it provides *insight* into our decision-making, helping us to know what to do with that information and compelling us to take meaningful action.

Data is only useful and relevant when it provides insight into our decision-making, helping us to know what to do with that information and compelling us to take meaningful action.

Until recently, we only had the ability to *collect* data, but we had no idea what to *do* with it. When the Fitbit was initially introduced, only 30 percent of users were still wearing them after three months. Why? Because the users had no idea what to do with the data. I remember when my husband first gave me a fitness-tracking wristband for Christmas. At the time, the thick orange band felt like an electronic inmate bracelet. However, I committed to trying the tracker out for a month, and was soon hooked. My favorite feature was actually the sleep tracker, which helped me see for the first time a visual confirmation of why I was always exhausted. Not only did I wake up to soothe a crying baby a couple of times a night, but it also took me a long time to go back to sleep each time, creating a horrible cycle that left me exhausted the next day. I dutifully consulted the bracelet's app to see what I should do about this information, and the overly cheerful user interface merely offered a lackluster tip that getting more sleep would help me to be more alert the next day. It's a wonder that I didn't break that band then and there. I also knew that this limited data set did not reveal the full picture. Was I anxious and fitful when I slept? Did I have good oxygen levels? How did my food choices the previous day affect my sleep? Did my sleep quality improve over time? There simply wasn't enough information to be useful, and I struggled to connect the dots. Fortunately, apps like Fitbit have improved dramatically in recent years as they have transitioned away from just *reporting* data to actually *interpreting* the data for us.

The next wave of technology and innovation is leading us toward actionable insight by aggregating data from our lives and then nudging positive behavior changes.[16] For instance, the "Addapp" app (say that ten times fast!) can pull data from multiple apps on my smartphone to make suggestions for my diet and exercise routine based on my past behavior. The app might recognize that my sleep quality has declined along with my activity level. However, if I were just able to take 2,000 more steps each day, I could significantly increase my chances of getting better sleep in the coming week. Now this is the kind of insight I can use!

The next wave of technology and innovation is leading us toward actionable insight by aggregating data from our lives and then nudging positive behavior changes.

Likewise, the Spire Stone is a wearable device that clips onto your waistband or bra to measure breathing patterns and to indicate whether you are in a state of calm, tension, or focus. The Spire Stone builds on the work of social psychologist Roy Baumeister, who once wrote that, "Good decision making is not a trait of the person. It's a state that fluctuates."[17] He went on to explain that the people with the best self-control actually structure their lives to conserve their willpower and reduce important decision-making during times of HALT (when you are Hungry, Angry, Lonely, or Tired). As a result, the Spire helps individuals learn to maximize their periods of focus by front-loading difficult decisions at the beginning of the day when they have energy and saving more routine or simple tasks for the late afternoon. This advice requires a bit of planning for your day, but can lead to dramatic improvements in your productivity levels without your having to work longer hours.[18]

Apps like these are part of an emerging field called persuasive technology, which explores how technology can persuade or "nudge" our behavior toward positive change. In the early 1990s, Stanford doctoral student BJ Fogg became fascinated with "how technology is . . . a channel for helping humans achieve their goals, for influencing people to do better, for changing organizations, and yes, for transforming the world."[19] He later went on to found the Stanford Persuasive Technology Lab, which currently studies a variety of applications of this tech, ranging from Amazon recommending to you products you might be interested in to Facebook recommending friends for you.

Other tech companies are catching on to this trend, using insights from psychology and behavioral economics to take on the role of Influencer in their app design. For instance, the corporate wellness training company Habit Design employed a team of game designers as well as PhDs to develop the first clinically tested, evidence-based platform to teach positive, sustainable habit development. Using online workshops, live peer coaching, and built-in incentives, Habit Design has managed to achieve an astonishing retention rate of 80 percent after three months (as opposed to traditional programs like seminars or counseling, that generally lose 80 percent of participants in the first ten days, according to Habit Design CEO Michael Kim).[20] Likewise, the fitness-tracking company Jawbone uses a virtual "Smart Coach" to send customized messages to users to help them increase their steps by 27 percent and improve their sleep by an additional twenty-three minutes per night.[21]

Persuasive technology is all around us, and it is literally changing how we think and how we develop habits. Of course, not all persuasive technology is aimed at moving you in the right direction. As technology produces more and more powerful consumer insights, savvy companies will use that information to target you to spend more money or to engage in more addictive behaviors. This type of marketing is so effective because it targets your microdecisions at moments when you are already scanning your environment for information—and it simply fills in customized suggestions to nudge your decision in a certain direction. Imagine the difficulty of sticking to your budget if stores could auto-sense your color and style preferences to morph the window display before you walk by. Or just try to stop drinking coffee if your new automated home kitchen knows that you love a warm cup of coffee in the morning and have never deviated from that pattern before.

Scenarios like these beckon the question: If companies are spending millions to gather data and insight into how you think, why aren't you using this treasure trove of information to get one step ahead? While persuasive technology continues to get better at "knowing you" through data-driven insight, the truth is that they only know *about* you from your past decisions—not who you want to be in the future. In his book *Persuasive Technology: Using Computers to Change What We Think and Do*, Fogg is careful to explain that technology should never be coercive; rather, we should always have an active choice in how, when, where, and why we let the digital world into our lives. (In Strategy #3, I will share with you a number of apps specifically designed to "nudge" you toward making more positive choices in your life.) By thinking smaller about how we use technology and information to shape our decisions, we become smarter about striving after our potential.

A HIGH-DEFINITION LOOK AT YOU

A few months ago I took my niece to the Perot Science Museum in Dallas. There was an amazing exhibit on nanotechnology in which you could examine a tiny butterfly wing under a gigantic microscope; to see greater detail, you simply spun a kid-sized wheel, which my niece did gleefully, spinning and spinning and spinning. And the image zoomed and zoomed and zoomed. There seemed to be no end to the zooming capabilities, as the microscope delved into the smallest particles of each cell. Truly, the depth and detail were mind-blowing! Although we were studying the tiniest of particles of an insect, the exhibit left me reeling as I contemplated just how complex and layered human nature is; by taking the time to think smaller about how our mind and body work together, we can uncover vast sources of insight.

If our minds could leave a print, like fingers do, what would that print look like? Each of us has a unique combination of identifiers in the way that we think and consequently act. Until recently, we could only make educated guesses about those identifiers. However, if I could wave a magic wand and produce a report that told you everything there was to know about your mind and body, would you even recognize yourself? Probably not, because we have never actually had this kind of information to provide that full mirror image of ourselves.

However, the confluence of the technological revolution and the cognitive revolution is giving us powerful new insight that we can use to strive toward our potential. For the first time in history, we can now peer behind the cognitive curtain to see how our minds work. By looking into our bodies and brains, we can actually gain a higher level of understanding of our unique *mind*prints, which include our thought processes, intentions, goals, and interests. And as we begin to pair new technologies with this greater awareness, we have the ability to know ourselves more deeply, shaping our actions and ultimately our future happiness as well in powerful ways.

The confluence of the technological revolution and the cognitive revolution is giving us powerful new insight that we can use to strive toward our potential.

One of the best ways to gain this knowledge of the intricacies of your daily life is through the use of wearables. Wearables come in all shapes and sizes today, from watches to clothing, shoe inserts to necklaces. Researchers from MIT have even created prototypes for temporary tattoos and even stylish nail art stickers that can function as interfaces for smartphones and other digital devices.[22,23] With the proliferation of new gadgets and wearables on the market, it's easy to get swept up in the possibilities for maximizing productivity, optimizing athletic potential, preventing injuries, being proactive about health decisions, even deciding when to bring up an idea to your boss or spouse depending on their mood. As I write this book, I am keenly aware that by the time it is published, it will be somewhat outdated by a proliferation of new tech, as there has already been a 44 percent increase in the number of wearables in the market in the last year. Rather than resist the onslaught, indulge with me for a moment and imagine what it would be like to quickly discover the answer to the following questions: Is your stress reaching critical levels? Are you drinking enough water? What's in your sweat? What are you feeling?

What if you understood yourself well enough to know when stress was mounting in your life before a migraine set in or a panic attack struck? Dr. Joel Ehrenkranz of the Intermountain Medical Center is helping to develop and test a new smartphone attachment for tracking cortisol levels at home, a procedure that normally costs upward of fifty dollars for just one sample and can take up to a week to produce results. Instead, Ehrenkranz's solution utilizes a small tube that costs less than five dollars to collect and analyze saliva samples in real time, producing lab-quality results in just five minutes. Tracking cortisol levels has far-reaching implications, as high cortisol levels affect the body's ability to recover from injuries, to fight infection, and to lose weight. For diabetic patients, cortisol levels are the key to regulating stress levels, an essential part of prevention and control of the disease. This simple app (not yet available to the public) could let us know when our stress levels might be reaching dangerous levels and suggest key times for us to take a breather or stop to meditate.[24]

What if you could look to see how much water you're actually consuming? Halo Wearables just released a wearable called the H1, which looks like a smartwatch but is used to monitor a user's hydration status. The H1 measures skin temperature, humidity, and air temperature to give users a relative value of their hydration using a green, yellow, and red index: Green zone lets users know tanks are full. Yellow zone informs users to consider fueling up soon. Red zone warns users tanks are empty.[25] For the athletes reading this, Kenzen ECHO has developed a small, flexible patch that uses a single bead of sweat to analyze critical biomarkers such as sodium and potassium to help improve performance and recovery, and to prevent injury.[26]

Or what if you could understand your emotions on a deeper level? The Mood Meter, designed through a partnership of researchers at Yale, is an app to build "emotional intelligence that lasts a lifetime." The app helps you to self-report your mood using the RULER program, for Recognizing, Understanding, Labeling, Expressing, and Regulating emotions. The app developers call this program a "gift of self-awareness for yourself, and for others."

Amazing, right?! Of course, if I actually wanted to use each of these technologies, not only would I have to spend a small fortune, but I also would have to invent a new skintight fanny pack to juggle all of these devices. But that situation will likely change, because wearables are just now emerging from their awkward adolescent years in the life cycle of innovation. Just as the giant car phones of the 1980s eventually morphed in form and function, so will wearables. We've already come so far, as can be seen from this humorous picture from 1993 of a group of MIT researchers decked out in some of the first wearables. Weighed down with large headsets, cumbersome jackets, and awkward belts, the researchers looked like a cross between a grunge band, the Ghostbusters, and a rogue army unit on a secret black-ops mission.

Steve Mann (pictured leftmost) founded the MIT Wearable Computing Project as its first member (http://wearcam.org/nn.htm). These early prototypes inspired a new field of research.

Yet these early gadgets served as the prototypes for some of the most sophisticated wearables in development today. At the far left of this picture is Steve Mann, who developed the first wearable augmented-reality system in 1974 and is widely known as the "father of the wearable computer." He is remotely connected to his camera for one of the first group "selfies."[27] Since this photo was taken, Mann has gone on to invent numerous breakthrough technologies, including Eye+Tap (a digital glass eye), MindMesh (a "thinking cap" that allows the user to plug various devices like a camera into his or her brain like a camera and use it as an "eye"), and HDR imaging (a technique where a camera takes three pictures at the same time with bright, medium, and dark lighting and then merges the images to give a greater range of shadows and highlights).[28,29]

In the course of writing this book, I had the honor and privilege of connecting with Mann over email as he discussed one of his latest project collaborations, called Meta, which offers the first commercially available pair of augmented-reality glasses.[30] At the steep price of $949, these glasses aren't in the budget for most of us, but their availability in the marketplace signals that augmented reality is actually a *reality*, not a futuristic technology. As Meta founder and CEO Meron Gribetz explains in his TED Talk, "In the next few years, humanity's going to go through a shift, I think. We're going to start putting an entire layer of digital information on the real world. Just imagine for a moment what this could mean for storytellers, for painters, for brain surgeons, for interior decorators, and maybe for all of us here today. And what I think we need to do as a community, is really try and make an effort to imagine how we can create this new reality in a way that extends the human experience, instead of gamifying our reality or cluttering it with digital information." As these innovations become tested, they will become more streamlined, stylish, and widely available in the same way that smartphones merged email, phones, and cameras. Patience, my friends. It's worth it.

The biggest innovations in society often come from very small changes in the way we go about our work or life.

The biggest innovations in society often come from very small changes in the way we go about our work or life. This past spring, I took a step into the near future. I was invited to join the XPRIZE Foundation as a Visioneer for one of its newest initiatives, a competition with a $10 million prize for the team that could come up with a solution to reduce the risk of chronic disease and improve the personal health of individuals across various demographics and around the globe. If this sounds far-fetched, keep in mind that previous XPRIZE competitions had led to successful ventures for privatized space travel, fuel-efficient car travel, and even rapid cleanup for oil spills.

But what was the personal-health crisis—was it physical, mental, or emotional? And was the health crisis actually "solvable"? With so many factors feeding into the crisis, how could one solution possibly make a difference? In the face of this massive goal, the solution that emerged was that we needed to think smaller about our health goals. Rather than create yet another large-scale government or corporate health initiative, we needed to empower individuals to make better choices on their own by giving them personalized information and insight. It's amazing how many decisions we make blindly—either due to lack of information or to the irrational hope that things will just even out eventually. For instance, we buy in to taking daily multivitamins without having any knowledge of which vitamins our unique bodies actually need. This is totally understandable because up until now there has been no way to obtain this information without having your blood drawn. But imagine in the not-so-distant future having a device like the futuristic Star Trek tricorder that you could sweep across your forehead (like a temporal thermometer) to get a read on which vitamins your body needs, and then using that information to inform your food choices throughout the day. So simple, yet so powerful. By providing in-depth, personalized insight, we can empower individuals with the information that matters to them in the moment, not after the fact, to bring about deep, lasting change.

As participants in the Digital Age, we can leverage technology like this to know ourselves better inside and out; and when we know ourselves better, we can begin to think smarter and make better choices about how to strive after our potential and ultimately find greater happiness in the future.

THINKING BEYOND TECH

Throughout this section, I have outlined all the ways that technology can help us understand ourselves and make better decisions. However, it's worth saying that technology is merely a tool to help us, not the answer to all of our problems.

> Technology is merely a tool to help us, not the answer to all of our problems.

You may be surprised to hear me, a self-described digital optimist, say that technology isn't the ultimate solution. But sometimes the most important part of thinking smaller is just stopping to *think*. Recently, my six-year-old daughter, Gabri, borrowed my iPhone and I overheard her asking Siri, "Who is God?" to which Siri promptly replied, "Fascinating question," and directed her to Wikipedia. This was such a small interaction, but also a striking one since this question is one of the oldest, most hotly debated questions of all time. And Siri had a reply in one second flat. Gabri would have been happy to accept Siri's matter-of-fact non-answer without thinking twice, but I took the opportunity to encourage Gabri to ask more questions. I explained that while technology is a fantastic tool for learning, we still have to round out this information with context and personal experience as well. Somewhere in my lengthy monologue, her eyes started to glaze over, but hopefully my point was heard.

Sometimes seeking to understand ourselves means tapping into the greatest supercomputer ever created, the human mind. This could mean stopping to unplug, look at the clouds, and just ponder some of life's greatest questions. It means connecting with our emotions, listening to our gut, or seeking wisdom through prayer or meditation. I'm sure some scientists may be frustrated when I write this, but I believe that humans are more than just a conglomeration of firing neurons. We feel things—we are not just machines acting on advanced algorithms, but we go beyond what the data tells us to discern what feels right. Because at the heart of knowing ourselves is a *heart*; it's what makes us human, better than machines, better than the tech that we create to know ourselves better. The ultimate act of knowing ourselves might be discerning when to think with our brains and when to feel with our hearts.

> At the heart of knowing ourselves is a *heart*; it's what makes us human, better than machines, better than the tech that we create to know ourselves better.

Our microdecisions define us as individuals. As we look toward the future of happiness, we might have technology at our fingertips, but what mindprint will we leave behind? Looking at the details of our own lives, we can see that the microdecisions we make give us powerful insights into how to make changes in our lives. Now it's time to use this information for transformation, by training our brains to use the principles of positive psychology to reach our full potential at work, at home, and in our personal lives.

SUMMARY

Self-knowledge is power—paying attention and giving intention to the microdecisions in our lives helps us to avoid limiting beliefs and make better choices for the future. New technology helps us to understand our own bodies and minds on an incredibly detailed level, enabling us to make better microdecisions about the future. However, we still have to tap into the greatest supercomputer ever created, the human mind.

Know thyself by:

- ✓ Learning to recognize limiting beliefs that might derail your best intentions
- ✓ Magnifying your microdecisions to understand where small changes can have big impacts
- ✓ Tracking progress in your life to determine where you have succeeded and where you have room for improvement

STRATEGY #3 TRAIN YOUR BRAIN

HOW TO PUT TOGETHER THE BUILDING BLOCKS OF A SMARTER, HAPPIER MIND

As a child, I was convinced that my dad just went to work to play. My father was a neuroscientist at Baylor University, and his office was full of brightly colored blocks to test intelligence, books by M. C. Escher to study perception, and even a soundproof room covered in blue foam that I thought was for gymnastics (but later learned was for studying brain waves during sleep). My favorite room in the office, though, was a computer lab down the hall labeled ~~“Work Room,”~~ “Play Room.” In this small closet of an office, there were three computer stations featuring the latest technology on the market. In 1985, when I was barely five years old (I'll let you do the math on my age), I remember sitting on my dad's lap as he taught me how to fill a screen on a Mac Classic with a fence-like pattern and then print it out on a space-age dot matrix printer. I was fascinated. Over the next few years, my dad would often bring us to his office in the summers while he taught classes, so we had ample time to get into mischief. Baylor did frequent systematic upgrades for tech, rotating the latest computer models and printers and software through that office for our entertainment. We literally believed that computers were just fun puzzles to solve, and we would explore every menu item and button until the screen froze, then turn off the computer and slink out of the room as if nothing had happened.

What we didn't know at that time was that we were part of a new generation growing up believing that computers were tools for exploration and adventure, and that our mindset of curiosity and exploration was actually a critical component in developing an ability to absorb information and innovate. As we look to the future, having a growth mindset about how technology can help us understand, routinize, and reinforce positive behaviors will undoubtedly be intertwined with our ability to strive after our potential.

No one understands this better than the faculty and students at the MIT Media Lab in Cambridge, MA, who aim to open the public's mind to coming possibilities. In the spring of 2016, I had the opportunity to visit the gorgeous museum-like Media Lab facility, which is like a modern-day digital playground. There's even an entire lab called Lifelong Kindergarten devoted "to developing new technologies that, in the spirit of the blocks and fingerpaint of kindergarten, expand the range of what people can design, create, and learn."[1] As I wandered from floor to floor observing the various exhibits, I felt like a kid in a digital candy store, eye-lusting over mannequins sporting the latest wearables, colorful orbs that displayed "the mood" of the stock market in real time, and the most gorgeous loaves of bread that were all made of "edible pixels." I even had a chance to try on a pair of one of the old Google Glass models, which I'm pretty sure made me look cross-eyed as I tried to read and respond to visual prompts in my upper right visual field (I can see where haters came up with the pejorative "Google Pain in the Ass" and why Google has since scrapped the project to rethink its design).

While at the Media Lab, I was able to meet with Javier Hernandez, a PhD student in the Affective Computing Group, to pick his brain about how technology will shape the future of happiness. Hernandez was part of a research team that developed the MIT Mood Meter, which was intended to be an artistic installation representing the mood of people on campus.[2] As individuals passed one of four Mood Meter locations on campus, they could see themselves on the large screen while a computer algorithm analyzed the intensity of their smile. Instantaneously, their image would appear as either a neutral face or smiley face. Data was then collected in real time over ten weeks to explore trends on questions such as, "Do midterms lower the mood?"; "Does warmer weather lead to happiness?"; and "Are people from one department happier than others?"

While smiles are only one barometer of happiness, the project aimed to create a "live portrait of a community, creating a time-changing location-based emotional footprint."[3] Now Hernandez is studying how biosensors can be used as a "communication prosthetic" to help better understand individuals who have communications impairments, such as those with severe cases of autism spectrum disorders.[4,5] By tracking an individual's response to stress and anxiety over time, parents and educators can begin to identify and anticipate triggers for tantrums or self-injury, which would be a major step forward for all. Wow!

While the digital playground can be awe-inspiring and fascinating, it also presents new challenges that we must learn to overcome. In the last strategy, we talked about how our decision-making can either derail us or drive us toward our potential, and we explored ways that technology could nudge our choices. The strategy in this section, however, delves more deeply into how we can use persuasive technology and positive psychology in tandem to elevate our mindsets and begin to benchmark change for sustainable growth. More specifically, I am going to share with you five skill sets you can use to actively raise your happiness level, and I will share with you some of my favorite gadgets, apps, and tools to help you do so.

THE CHALLENGE:

DEALING WITH A BRAIN THAT HAS A MIND OF ITS OWN

Untrained brains are a bit like puppies, developing "a mind of their own" and causing a cascade of consequences. For those of you who have ever raised a puppy (or a child, for that matter), you will understand how shocking it can be when your adorable angel first gets that glint in his or her eye. I remember the morning that this happened to me about a month after adopting a new puppy into our family. I was of course running late for work, and my puppy was taking an epic morning stroll to find the perfect spot to do her business. The moment that she finished, I swooped in impatiently to pick her up, and would you believe that my cuddly lump of fur looked me square in the eye and did a side-lunge-juke to evade me? Not only that, but I watched in horror as she squeezed through my fence and dashed into my neighbor's muddy garden with the joyful bound of a gazelle. I chased after her; I scolded her; I used my high-pitch-fake-happy voice; I even tried to trick her into coming with a treat. But in that moment, I realized with chagrin that I had never bothered to teach my puppy the all-important recall command "come," as in, "come here right now, darn it!" Assuming that she would always be a pliable lump of snuggly fur, I had underestimated my puppy's growing mind and the need for attention training.

Likewise, we fail to train our brains to "come" when called, assuming that our minds operate on autopilot and will act in our best interest. Yet, as we all know from personal experience, when challenges come, our bodies aren't always well trained to respond on command. Sometimes our bodies take over, resorting to a "fight or flight" response, and instead of behaving in our best interest, our mind turns into a mischievous gazelle-puppy-on-the-run. Eknath Easwaran, one of the twentieth century's great spiritual teachers, once humorously explained the dissonance between our bodies and minds in his book *Meditation*:

> *Suppose I come out one morning, start up my car, and drive off to give a talk on meditation in Milpitas, south of San Francisco. As soon as I cross the Golden Gate Bridge, my car veers east toward Interstate 80. I keep trying to turn the wheel, but there is tremendous resistance—the steering mechanism is ignoring me. "Milpitas!" I protest. "We're supposed to be going to Milpitas!" But the car only roars insolently, "Reno! Reno! We're going to Reno!" Then I think I hear it snicker, "Why not sit back and enjoy the ride?" Would we put up with that? Well no, not from our cars. But most of us do from our minds. In theory we would like the mind to listen to us obediently, but in fact it will not—chiefly because we have never taught it how.*[6]

We haven't taught our minds how to listen to us obediently because we either didn't even know it was possible or had no idea how to do so.

We haven't taught our minds how to listen to us obediently because we either didn't even know it was possible or had no idea how to do so. Fortunately, the last two decades of research in the field of positive psychology have revealed that training our brains is not only possible, but that doing so can actually change the shape and function of our brains by improving neural plasticity (you can, in fact, teach an old dog new tricks), increasing gray matter (the density of brain cells that drive how fast you can move, learn, and sense things around you), and strengthening neural networks (the pathways for our brain to talk to itself and the rest of the body).[7,8] The best part is that you don't have to be a neuroscientist to start training your brain, nor do you need sophisticated equipment. In the next few pages, I will share with you several low- or no-cost resources to get started. It's time to start training our minds now—and just like with a puppy, the sooner the better.

THE STRATEGY:
OPTIMIZING YOUR MINDSET

If happiness is the answer that we seek, then optimizing our mindset is the algorithm to get us there.

If happiness is the answer that we seek, then optimizing our mindset is the algorithm to get us there. In psychology, the term *metacognition* is used to describe this optimization of the mindset, simply as an awareness of how one thinks. A recent study of mindfulness in the workplace found that the ability to step back from automatic, habitual reactions (those fight-or-flight responses) is highly predictive of work engagement and well-being.[9] For instance, if you often feel your blood starting to boil when you hear a colleague down the hall talking too loudly on the phone, metacognition and mindfulness give you the power to choose a different reaction—perhaps taking a deep breath, using the opportunity to go for a walk, or listening to your favorite music. Mindfulness also creates positive job-related affect, higher levels of engagement, and increased psychological capital (hope, optimism, resiliency, and self-efficacy), all of which are highly predictive of success at work. Specifically, a positive and engaged brain is 31 percent more productive, three times more creative, and ten times more engaged.[10]

A positive and engaged brain is 31 percent more productive, three times more creative, and ten times more engaged.

To understand why mindset is so important, imagine yourself in a blizzard with snow coming at you in all directions. Each snowflake you see is a tiny bit of information swirling around you, and at any given second, there are eleven million snowflakes surrounding you. But the human brain is a single processor, meaning that at any given time, it can only focus on forty bits of information. That's forty snowflakes in a blinding snowstorm. And which information does the brain focus on? What GoodThink and other positive psychology researchers have found is the more that your brain scans the world for stresses, hassles, complaints, and threats, the less your brain is able to scan the world for the positive things that we know help drive us toward our potential (remember, the Greeks defined happiness as the joy we feel striving toward our potential).

Stresses, hassles, complaints, and threats all fall into the negativity category, what Carol Dweck calls the foundation of "fixed mindset."[11] In her book *Mindset: The New Psychology of Success*, Dweck explains that, when contemplating new tasks or challenges, individuals with a fixed mindset will say things like, "I'm not good at math"; "I'm not a social person"; or "I'm just not that funny." These individuals mistakenly believe that their innate talents determine their success in life, as though success is a foregone conclusion. Let me tell you, if my oldest daughter tried to tell me that she couldn't help with dishes because she wasn't good at it, I would most likely say, "Suck it up, cupcake. It's time to learn." We don't accept fixed mindsets in our children because they are young and we expect them to grow. However, as we get older, we justify our attitudes about life, using our past experiences to unconsciously harbor a fixed mindset, even if the attitudes are unfounded and unhelpful.

Optimism, on the other hand, fuels a "growth mindset," which Dweck defines as the belief that our most basic abilities can be developed through dedication and hard work. By scanning the world for the positive, we can begin to transform our past failures, hurts, and fears into a source of potential growth—a process that paves the way for long-term happiness. In fact, researcher Barbara Frederickson discovered happy people often experience an "upward spiral" when they feel happy: Happy people like to develop new skills, which leads to new successes, which results in more happiness, which causes the process to repeat itself.[12] I want to be on that upward spiral! But of course, it is easier said than done.

USING TRANSFORMATIVE TECHNOLOGIES FOR BRAIN TRAINING

Believing that your behavior matters is at the heart of training your brain.

Believing that your behavior matters is at the heart of training your brain. While this idea is not new—numerous scientists, religious gurus, and thought leaders have preceded me in espousing this idea—the Digital Era has opened a new frontier of understanding how we can get *strategic* about brain training by using technology to reinforce positive behaviors in our lives. I recently interviewed Jen Moss, co-founder of Plasticity Labs and author of *Unlocking Happiness at Work*, who is at the forefront of this field, to learn more about her approach[13]:

Amy: Why did you decide to start a company focused on brain training?

Jen: My husband, Jim, and I started on our mission to give one billion people the tools to live a happier and higher-performing life after Jim became acutely paralyzed in September 2009. We joke that we're accidental entrepreneurs, and yet, it's highly accurate. Before Jim became ill, he was a gold-medal-winning professional Canadian lacrosse player. But when he contracted a rare illness that rendered him unable to walk, he was forced to face a new challenge—relearning how to be high-performing in a new life without pro sports. We both figured it out together because we had to. We had one child and another on the way, and, as parents, you have a responsibility to remain positive and functioning for your children.

We took many lessons away from this experience over the years, but one very important aspect of Jim's recovery struck us as significant. Jim's doctors said that he may never walk again, or if he did manage to walk, it would be with assistance. But Jim walked out of the hospital after only six weeks—a few days short of the birth of our second child. What we learned was that Jim, through his immersion in the world of sports, and sports psychology, had developed resiliency, optimism, growth mindset, gratitude, hope, and the list of high-performing traits goes on. This psychological fitness training would help him in his healing process, and the knock-on effect would be our ability to turn post-traumatic stress into post-traumatic growth.

Amy: Why did you choose to focus on a technological solution to raising happiness and performance?

Jen: The marriage of technology and happiness may seem like a strange partnership to some, but it is actually quite critical to our mission. The ability to scale communication through digital media was beyond anything within our capacity even twenty years ago. From carrier pigeons to instant messaging, we've seen some pretty big advances in communication capacity over the last 150 years. But, it might be digital social collaboration that will go down in history as one of the biggest disrupters to our communication norms in the last century—and it's one that is both beneficial and detrimental to our health. We've been able to learn from the Plasticity Labs data that positive interventions delivered online can increase mental and physical well-being and overall happiness by 30 percent in just ten days. Imagine now we could deliver that kind of ten-day intervention to anyone with a mobile device or a laptop. The speed in which we can communicate and assimilate knowledge is faster than ever before.

But insofar as technology is rapidly moving and it offers us the benefit of scale, it also poses a threat of becoming too reliant on it for our happiness. Jim and I are incredibly cognizant of that and have built offline activities in the app to remind us that it's time to step away from the tech. Seems ironic that our app reminds us to stop using our tech, to get up and stretch, or take a walking meeting. But it's a crucial aspect to authentic happiness—moderation.

Amy: Given your work, you are steeped in technology on a daily basis. How do you go about training your brain while balancing a sense of well-being in your personal life?

Jen: Jim and I try to apply these same principles in our family life. We eat dinner together every night without any TVs on in the background or tech at the table. We don't even answer phone calls. Every person shares their gratitudes while we eat and it creates some pretty amazing dialogue about everyone's day. Our kids are fluent in gratitude because we intentionally practice it daily, and since they know they have to report to mom and dad what they are grateful for at dinner, they tend to seek it out during those otherwise neutral moments in their lives. They are also involved in competitive dance, and for those of you who know how much time that takes, it pretty much squashes any concern of technology overexposure. However, we definitely have fun with tech, too. Jim and my son go out Pokémon-hunting quite often through our neighborhood and they have a blast sharing all their awesome catches!

My conversation with Jen reminded me that authentic and lasting behavior change is not a sterile process of tracking and improving metrics, but rather is a very personal process of sifting through your real-life experiences, fears, biases, and hopes to find the best path forward. True transformation comes when we connect brain change with heart change. If we truly want to strive toward our potential, we need all the help we can get—and transformative technologies like Plasticity Labs can be a tremendous resource to facilitate growth and provide accountability.

True transformation comes when we connect brain change with heart change.

GETTING YOUR SKIN IN THE GAME

Over the course of writing this book, I have had the opportunity to test-drive a number of new technologies aimed at increasing well-being. I got my skin in the game. I soon realized that I was going to need to add about an hour to my morning routine to "suit up" each day with my various wearables. And of course, I would need to add another hour to my bedtime routine to download my results and track my progress. I could literally turn into a walking robot of wearables to increase my awareness . . . or as one of my dear friends says, I could just do yoga. And she has a great point. Why do we need all these gadgets to increase our awareness? If we really tuned into ourselves, shouldn't we know exactly how we are feeling?

The goal of brain training is not to replace your brain—it's to augment your focus and awareness of what your body has already been trying to tell you all along. For instance, my back frequently hurts after I've been sitting for long hours at my desk. I know I don't have the best posture, but for the life of me, my mind slips and I start to slouch again. I decided to try two different devices designed to improve posture: Lumo-Lift, a tiny magnetic square that attaches to your shirt or bra strap, and Upright, a wearable device that attaches to your back with a light adhesive and Velcro.[14,15] Both vibrated gently when my posture shifted out of alignment to remind me to sit up straighter, which my devices seemed to do nonstop for the first few minutes. However, I did eventually learn from my mistakes. Within three days, I was able to go for long stretches without being reminded to sit taller. These wearables helped me to increase my awareness, strengthen my muscles, and create muscle memory in my brain by forging new neural pathways. While the goal of these devices is to create a sustained new habit, ideally these wearables are just short-term interventions for long-term gains.

Now, my friend would argue back, "But, Amy, you can do yoga to strengthen your back and increase your awareness," about which she is completely correct. But let's be honest, I've known about the benefits of yoga for years, but I haven't consistently followed through on practicing it. Sometimes, what works for one person may not be the best for someone else. Carol Dweck explains: "Students need to try new strategies and seek input from others when they're stuck. They need this repertoire of approaches—not just sheer effort—to learn and improve."[16] This might mean that if my goal is to improve my stance, I might need to do yoga and technology-assisted posture training.

As you begin to contemplate the domains in which you want to train your brain, the key is to target a new behavior in your life that you are excited about and then run after it with a passionate intentionality, using all available resources at your disposal.

SETTING THE STAGE FOR SUSTAINABLE POSITIVE CHANGE

At GoodThink, we get asked all the time how to make positive habit change "stick" in our lives. So I was particularly excited when I met Ofer Leidner, co-founder of Happify, in 2012. Strategically located across the street from the delicious Dean & Deluca grocery store in New York (no wonder they are happy), Happify's office possesses a distinctly start-up feel (complete with a small putting green). The conference room where we met featured a combination of antique and modern phones, lending a comfortable charm to an organization at the forefront of cognitive brain training. Leidner explained to me that Happify is a website and mobile app that helps people improve their emotional well-being, cultivate resilience, increase mindfulness, and create lasting happiness through research-backed activities and games.[17]

Happify is aimed at cognitive brain training, but specifically focused on providing evidence-based solutions for better emotional health. Although many people believe that their genes or the external environment determines their happiness, it turns out that roughly 40 percent of our happiness is controlled by our thoughts and actions, 10 percent by our circumstances, and the other half is genetic.[18] It's hard to change our genetics and circumstances, but we can train our brains to be happier.

It's hard to change your genetics and circumstances, but you can train your brain to be happier.

Leidner teamed up with his long-time business partner Tomer Ben-Kiki to create a platform to develop and study how key skill sets could raise an individual's happiness levels for the long run. Here are a few ways that Happify helps to train the brain:

- ✓ Need help focusing on the positive? A game called "Uplift" teaches your brain to scan the environment for the positive, thereby improving your mood and reducing negative thinking. As hot-air balloons float by, click on words like "joy" or "radiant" while ignoring words like "criticize" or "angry."
- ✓ Searching for a way to relax? Choose the "Serenity Scene" activity. Perfect for someone feeling overwhelmed with a long to-do list, these guided relaxation tracks can help people unwind, feel less anxious, and get a fresh charge of energy (grounded in brain-scan research).
- ✓ Want games for your children to try? "Negative Knockout" is an Angry Birds–like game where you use a slingshot to destroy words that describe your biggest challenges that day.

Two months after regularly using the platform, 86 percent of users report feeling significantly happier. This impressive statistic highlights how technology can create positive change in our lives, enabling us to rise above our genes and environment to tap into our greater potential.

THE FIVE TARGET SKILL SETS FOR ELEVATING YOUR MINDSET

In the same way that you might go to the gym to exercise different muscle groups, so you can intentionally develop different skill sets that improve your overall sense of well-being and happiness. While there are a number of different positive habits that you can develop, I love Happify's S.T.A.G.E. framework for developing skill sets that elevate your mindset: Savor, Thank, Aspire, Give, and Empathize.

Savor

> Savoring is a quick and easy way to boost optimism and reduce stress and negative emotions. It's the practice of being mindful and noticing the good stuff around you, taking the extra time to prolong and intensify your enjoyment of the moment, making a pleasurable experience last for as long as possible. So whether it's preparing a meal, pausing to admire the sunset, or telling a friend your good news—the idea is to linger, take it in, and enjoy the experience. Eventually it'll become a habit—one you'll never want to break. Research by Fred Bryant, a professor at Loyola University Chicago who coined the term "savoring," shows that those who regularly and frequently savor are happier, more optimistic, and more satisfied with life. Bryant describes savoring as threefold, meaning we can savor the past (by reminiscing), savor the future (through positive anticipation), or savor the present (by practicing mindfulness).

Savor: The practice of being mindful and noticing the good stuff around you, taking the extra time to prolong and intensify your enjoyment of the moment, making a pleasurable experience last for as long as possible.

There are many savoring techniques—and you may find that you gravitate toward some, but not others. In addition to the Happify platform, whose activities and games help you build all five of the happiness skills listed below, here are a few of my favorite apps and devices to help you practice the art of savoring:

Apps and Devices to Practice the Art of Savoring	
Headspace	Headspace is a "gym membership for the mind." This app delivers multiple offerings for guided meditation.
Remindfulness	The Remindfulness app delivers gentle reminders to help you stay mindful throughout your busy day.
Muse Headband	Muse is a wearable brain-sensing headband that can measure the wearer's level of calm. The goal of Muse is to use biofeedback to train your brain.
Insight Timer	The Insight Timer app combines a meditation timer with a mindfulness guide. Enjoy guided meditations and find other people in your area who are meditating.
Mindfulness Training App	The Mindfulness Training App walks you through a number of meditation practices and styles to cultivate spiritual awareness and physical well-being.

Thank

The simple act of identifying and then appreciating the things people do for us is a modern-day wonder drug.[19] It fills us with optimism and self-confidence to know that others are there for us. It dampens our desires for "more" of everything—and it deepens our relationships with loved ones. And when we express our gratitude to someone, we get kindness and gratitude in return. In studies led by Martin Seligman, the father of positive psychology, people have written gratitude letters to someone they've never properly thanked, and seen immediate increases in happiness and decreases in depressive symptoms. Bob Emmons, professor of psychology at the University of California, Davis, is a leading researcher in the field of gratitude and author of *Thanks!: How the New Science of Gratitude Can Make You Happier.*[20] He believes everyone should try practicing gratitude because the benefits are so powerful: "First, the practice of gratitude can increase happiness levels by around 25 percent. Second, this is not hard to achieve. A few hours writing a gratitude journal over three weeks can create an effect that lasts six months, if not more. Third, cultivating gratitude brings other health effects, such as longer and better quality sleep time."[21]

Thank: Identifying and appreciating the things people do for us fills us with optimism and self-confidence.

It's worth noting that there are a number of apps to help you develop and track your practice of gratitude, but verbal and physical thank-yous are equally—if not more—effective:

Apps for Being Thankful	
Gratitude Journal	The Gratitude Journal app is a completely private way to record your gratitudes with words and images.
Gratitude 365	Gratitude 365 is a beautiful and easy way to write in your gratitude journal.
Happier	Happier is a fun social-gratitude journal combined with a positive community. Be inspired to keep track of the small happy moments in your day.

Aspire

Feeling hopeful, having a sense of purpose, being optimistic. Study after study shows that people who have created meaning in their lives are happier and more satisfied with their lives.[22] You, too, can feel more upbeat about your future and your potential. And who doesn't want that? Genuine optimism is a friend magnet. It also makes your goals seem attainable and your challenges easier to overcome. Bottom line: you'll not only *feel* more successful, you'll *be* more successful. A person's level of hope is shown to correlate with how well he or she performs tasks. Using one's strengths in daily life, studies have found, curbs stress and increases self-esteem and vitality. Another study found that participants who were asked to imagine their future in an optimistic light increased their levels of happiness over the next six months. Believing that your goals are within reach promotes a sense of meaning and purpose in life—a key ingredient of happiness.

Aspire: Feeling hopeful, having a sense of purpose, and being optimistic.

Here are some fun apps and platforms for helping you strive toward your potential:

Apps for Aspiration	
Live Happy	A magazine, website, and resource about a timeless quest: living a happy life.
Live Intentionally	The Live Intentionally app brings conscious living to the forefront of your life by helping you set and track your intentions for self-improvement.

Plasticity Labs	Plasticity Labs is a workplace social platform designed to help measure and build organizational happiness.
Potentia Labs	Online platform offering courses for organizations to engage their employees with scientifically proven techniques to help them build the social, emotional, and professional habits they need to overcome obstacles, big and small. The step-by-step approach guides students through the experiences, helping them practice what they're learning along the way, in just three minutes a day.

Give

Everything about giving is a no-brainer. Obviously, when you give someone something, you make him or her happier. But what you might not know is that the giver—not the receiver—reaps even more benefits.[23] Numerous studies show that being kind not only makes us feel less stressed, isolated, and angry, but it also makes us feel considerably happier, more connected with the world, and more open to new experiences. In one famous study, Sonja Lyubomirsky asked students to commit five random acts of kindness each week for six weeks. Whereas the control group (who did not do acts of kindness) experienced a reduction in well-being, those who engaged in acts of kindness showed a 42 percent increase in happiness. A twist on this study found that we're also happier when we spend money on other people than when we spend money on ourselves. And a 2006 study found that simply reflecting on nice things we've done for other people can actually lift our mood. Stephen Post, a bioethicist at Case Western Reserve University and founder of the Institute for Research on Unlimited Love, is a pioneer in the study of altruism and compassion. His research shows that when we give of ourselves, everything from life satisfaction to self-realization and physical health is significantly affected. Mortality is delayed. Depression is reduced. Well-being and good fortune are increased.

This might just be my favorite list of apps for developing a spirit of giving in your life. Thanks to leveraged partnerships and crowdsourcing, the world of giving is experiencing a revolution in the way that small gifts of time, talent, or treasure can have big effects.

Give: Being kind not only makes us feel less stressed, isolated, and angry, but it also makes us feel considerably happier, more connected with the world, and more open to new experiences.

Here are my favorites:

Apps for Generosity	
Pay It Forward	The Pay It Forward app aims to spread random acts of kindness. The app will suggest a "random deed," such as giving up your seat on a crowded train or sharing an inspirational quote.
Ripil	Track and share the good you create in the world or log the deed for yourself and use Ripil as your personal kindness journal.
Deedtags	Deedtags is an app that gives you daily good-deed challenges.
BeHppy	BeHppy is an anonymous network that facilitates and promotes happiness. The smiles and replies you receive from the community become coins toward a charitable donation.

Charity Miles	Charity Miles donates ten cents to your favorite nonprofit for every mile you bike and twenty-five cents for every mile you run.
One Today	The One Today app by Google uses microphilanthropy to bring exposure to different nonprofits every day, suggesting that users donate $1 to their favorite cause and amplify the impact by matching their friends' donations.
Feedie	Use the Feedie app to post a photo of your food on social media, and participating restaurants will make a donation to help feed orphaned and at-risk schoolchildren in South Africa.
Instead	Much like One Today, the Instead app encourages small donations that make a big impact. Suggesting increments of $3 or $5, Instead will inspire you to give back without breaking the bank.
GiveGab	With GiveGab, you can connect with friends to make a difference in your community. Find great local volunteer opportunities, see where others are volunteering, and share your experiences.
Norm-Social Philanthropy	Discover volunteer opportunities and become an advocate for the charities you love.
Spare	Round up your dining-out bills to the nearest dollar to fight hunger in your own city.

GiveMob	GiveMob is a charitable-giving app for iPhone that allows users to donate small sums of money ($5-$10) to featured charities through SMS.
Impossible	Join a social-giving network that shows your requests to the people who may be able to grant them and shows you requests you may be able to grant.
Donate a Photo	For every photo you share through Donate a Photo, Johnson & Johnson gives $1 to a cause you want to help.

Empathize

Empathy is a powerful word packed with lots of different interpretations. It's the ability to care about others. It's the ability to imagine and understand the thoughts, behaviors, or ideas of others, including those different from us. If you care about the relationships in your life—and who doesn't?—learning the skill of empathy has enormous payoffs. When we empathize with people, we become less judgmental, less frustrated, less angry, or less disappointed—and we develop patience. We also solidify the bonds with those closest to us. And when we really listen to the points of view of others, they're very likely to listen to ours. Strong relationships are essential to happiness, according to Ed Diener and Martin Seligman, and practicing empathy will go far in nurturing the relationships in your life. Richard Davidson, professor of psychology at the University of Wisconsin-Madison, was the first to show that compassion is a skill that we can all learn because the brain is constantly changing in response to environmental factors. Likewise, self-compassion is also teachable. Research by Kristin Neff, a pioneer in the field, suggests that people who have more self-compassion lead healthier, more productive lives than those who are self-critical.

Empathize: The ability to care about others, to imagine and understand the thoughts, behaviors, or ideas of others, including those different from ourselves.

To explore apps that help foster a sense of empathy, check these out:

Apps for Empathy	
Moodies	Moodies is an app that can recognize different emotions in people's speech. The developers envision that this app can help individuals understand their own emotions, decode others' emotions, help track emotion over time, and even potentially detect when an individual is lying.
SoulPancake	Founded by actor Rainn Wilson (Dwight in NBC's *The Office*), SoulPancake is a place where people come and talk about stuff that really matters, such as spirituality, religion, death, love, life's purpose, creativity, and anything else you can think of.
Humans of New York	This app showcases a catalogue of photos along with the stories behind New York City inhabitants with the intent of raising a greater sense of empathy around the world.
Campfire	Designed to strengthen social networks, the Campfire app encourages you and your friends to put your phones down to develop real-world connections. While you interact, your phones play a game, earning points by the length of time that you keep your phone down.

SUSTAINING POSITIVE CHANGE

In *The Happiness Advantage*, Shawn describes how the legendary swordsman Zorro famously taught a young and impetuous pupil to master the sword by first training in a small circle. Only after gaining control of that small circle could the student expand his circle and move on to other, larger skills, like swinging from ropes and chandeliers. Given that *Zorro* made its debut in the era of black-and-white television, how would Zorro handle a pupil in the Digital Age who is distracted by limiting beliefs, weighed down with assorted devices, and obsessed with the swing-analyzing sensor on the bottom of his sword? While the tools may be different, I imagine his basic strategy for taking on the world would actually be the same now as it was then.

Happy Hacks to Get You Started

1. **Start small.** As tempting as it might be to try out every new app, wearable, and gadget that you can get your hands on, the most effective way to create sustainable positive change in your life is to start small. Using the principle of the Zorro Circle, I recommend that you select one skill set that you want to cultivate, and practice that skill for at least twenty-one days (the minimum length of time that it takes to develop a habit). Set small manageable goals, and increase the difficulty level as you master each level. And don't be surprised if you find you need more time to set your habit; researchers like Dr. Christine Carter at the Greater Good Science Center maintain that many of us need more like ninety days to set a firm habit.
2. **Select the right tool.** It's worth noting that Zorro did not start his pupil out by learning to fight with a sledgehammer; he very carefully selected a tool suited to creating small, incremental changes. Likewise, if we want to create change in our lives, we need to find the right tool to do so. This might mean testing out a few different apps and then homing in on the one that you will actually commit to using.

3. **Know when to move on.** One of Zorro's strongest tactics in battle was to know when to move on to higher ground and try a new position. While many developers would love for their technologies to become an integral part of our everyday lives, I prefer to think of habit-developing technologies more as short-term interventions. I love that there are wearables to help me improve my posture, but I don't really want to wear that device for the rest of my life. My motto is: make a change, sustain the change, and then move on to the next change.

LEADERS TRAIN THEIR BRAINS

Training your brain is not just a hobby for overachievers; it's a leadership strategy. Whether you are a CEO, summer intern, corporate employee, graduate student, athlete, or even a parent, these training skill sets are the building blocks of positive habit change in your life. I could of course make the argument that training your brain is ultimately self-seeking, reinforcing a desire for self-preservation and enhancement. But there is a collective argument for brain training as well.

Whether you are part of a business, a team, or a family, the whole is only as good/productive/successful/happy as the parts that make it up. I promise you that once you commit to training your brain in the ways that I outline above, you will find that not only are you happier, but that the larger community around you is also changed for the better. This happens because we are all wirelessly connected.

Once you commit to training your brain in the ways that I outline above, you will find that not only are you happier, but that the larger community around you is also changed for the better.

Yes, in this book about being wired, plugged in, and positively tech-crazy, I'm suggesting that our strongest connections to each other are actually wireless. If you don't believe me, try this one-minute experiment: the next time you're waiting in line at Starbucks, start to look at your wristwatch nervously and sigh repeatedly and loudly. Within seconds, witness more than half of the people in line with you begin to mimic your stressed-out antics.[24] Yes, you should feel bad. Now go do something good. At your next office meeting, walk in with a big smile on your face, making sure to greet each person with that warm, heartfelt smile. I guarantee you will change the entire mood of that room. This superpower is fueled by the mirror neurons in our brains, connecting us wirelessly to each other and giving us the subconscious (and conscious) ability to spread happiness. If you train your brain to focus on the positive, you will begin consciously (and then hopefully subconsciously) to spread positivity to those around you.

If you train your brain to focus on the positive, you will begin consciously (and then hopefully subconsciously) to spread positivity to those around you.

At GoodThink, we talk often about how happiness is a personal choice, and it's one that is independent of our genes and our environment. That being said, there are tools around us that can bolster our efforts, and there are ways that we can influence our environments to make that choice easier. In the next section, I will share some of the ways that you can set yourself up for greater happiness in the home, in the workplace, and in the places you learn and grow.

SUMMARY

Recent advances in positive psychology reveal that we can train our brains to improve happiness and performance by using the S.T.A.G.E. framework. To create sustainable positive change in our lives, the key is to identify a target skill set, home in on one habit, assess your progress, and make the change "sticky" by setting simple, relevant, and realistic goals.

Train your brain by:

- ✓ Developing an optimistic mindset to fuel your growth
- ✓ Using the S.T.A.G.E. framework (Savor, Thank, Aspire, Give, and Empathize) to learn skills for improving your mindset
- ✓ Tapping into technology to bolster your success and track your progress

STRATEGY #4 CREATE A HABITAT FOR HAPPINESS

HOW TO BUILD GREATER HAPPINESS INTO OUR HOMES, WORKPLACES, AND COMMUNITIES

If you ask my husband and me why we bought our current home, you will get a unanimous and resounding answer: the beanbags. Admittedly, this was not the *only* reason we bought the house, but after looking at close to 100 homes in the Dallas area, we fell in love and made the offer, on one condition: the beanbags must stay. What was so special about these beanbags? Quite simply, the size and comfort. Imagine taking a nap on the soft, warm belly of a big black grizzly bear—without the lethal paws and appetite for destruction. The moment our children spotted the bags in the media room of the prospective house, they became Olympic gymnasts, back- and forward-flipping into them at breakneck speed. (Like any good parent, I had to personally test the safety of these stunts to ensure quality control.)

Upon moving into the house, I insisted that we remove the covers of our new plush furniture to wash off lint and dog hair left over from the previous owners, thinking this to be a prudent and parentally responsible decision on my part. I carefully air-dried the covers so that they wouldn't shrink in the dryer, and then happily started to slip them back onto the grizzly-size bags. I made two miscalculations. First, this was no one-person job. So I recruited my hubby's help even though he resisted (hmmm . . . had he perhaps suggested that I not remove and wash the covers because they would be a pain to put back on?). The second miscalculation was an affirmation of my husband's fear. I did not anticipate that my docile grizzly-bear beanbags were missing teeth; or rather, the *zippers* on the covers were missing a couple of teeth here and there, making it virtually impossible to get the zip to catch and lock together. We pulled and tugged, we jumped and shoved, but every time we made progress, the zipper would burst forth triumphantly. I couldn't back down . . . I had committed to restoring these beanbags, the children could barely contain their excitement about playing on them, so gosh darn it, I was going to make it work. Frustrated, sweaty, and grumpy, my husband and I finally took a break to recruit our six-foot-four, ex-football-player neighbor as reinforcement. One hour later, our trio successfully conquered the zippers and cautiously plopped onto the bags in exhaustion.

The following sections offer guidance about how you, too, can create your own grizzly bear–sized den of happiness.

THE CHALLENGE:
DECIDING IF TECHNOLOGY IS OUR FRIEND OR FOE

While you might not have to wrestle with a giant, bear-sized beanbag anytime soon, the struggle isn't unlike the one we face each day when it comes to successfully integrating the steady stream of technology into our lives, without that stream taking over and controlling us.

Often we feel obligated to "make it work" with each advancement in tech—perhaps because of an initial investment, perhaps because an employer demands it to be so, or perhaps even out of an unfounded fear that if we do not adopt said tech, we will no longer be current (and therefore no longer relevant). However, just because an item or advancement is new and useful doesn't mean that it is new and useful for each of us personally. In fact, some tech can have a predatory effect upon our happiness, throwing off our delicate balance of productivity and well-being.

Consider the effect a new predator can have on the environment. In the last decade, wildlife in the Florida Everglades has been under constant threat from Burmese pythons, non-native predators that are thought to have been introduced to the area through the exotic-pet trade.[1] These twenty-three-foot, 200-pound snakes are known as apex predators because they are at the top of the food chain and can eat everything from bunnies to gators. As documented throughout the world, by simply introducing just ONE invasive species into a habitat, the result can be devastating for the entire ecosystem.

However, just because an item or advancement is new and useful doesn't mean that it is new and useful for each of us personally. Some tech can have a predatory effect upon our happiness system, throwing off our delicate balance of productivity and well-being.

However, when we apply this metaphor of technology to invasive species, I argue that it is not the tech itself that is the predator; rather, it is our mindset and choices that are the threat. As William Shakespeare wrote, "There is nothing either good or bad, but thinking makes it so."[2] We know from Strategy #3 that tech has the potential to add significant value to our lives, expanding our minds, fueling our growth, and fostering community. But we also know from our day-to-day lives that some technologies are life-sucking and others can be life-giving (and one tech can be both for different types of people!). Rather than trying to carve out a tech-free sanctuary in our lives, we want to avoid introducing things into our environment that prey upon our happiness and are out of alignment with our larger goals.

> Rather than trying to carve out a tech-free sanctuary in our lives, we want to avoid introducing things into our environment that prey upon our happiness and are out of alignment with our larger goals.

Strategy #4 illuminates how you can create in your life a habitat for happiness, which I define as an ecosystem that allows all parts of your life—digital and otherwise—to coexist and thrive. By pulling together research from the fields of positive psychology, physics, education, and design, I'll share with you ideas for rethinking the spaces and places in which you live, work, and learn.

THE STRATEGY:

RESHAPING SPACES, PLACES, AND THE FENCES IN BETWEEN

MAKING SPACE FOR FUTURE HAPPINESS

In 1965, Gordon Moore, co-founder of Intel, stated what is now called Moore's Law: that the number and speed of microchips will double every two years or so. For the past fifty years, this rule of thumb has driven the computing industry. However, in March 2016, the worldwide semiconductor industry formally acknowledged that Moore's Law was nearing its end. Instead, computing will turn to what is being called the "More than Moore Strategy": Rather than making microchips better and letting the applications follow, the industry will start with applications and see what chips are needed to support them.[3] Essentially, this means that the proliferation of commercial technology is about to reach exponential proportions. In fact, the wearable-device market is expected to grow 44 percent by the end of 2016 and continue growing at a 28 percent compound annual growth rate over the next five years.[4] Thank goodness I'm getting this book turned in before the end of 2016!

What does this mean for you and me? It means that we need to create spaces in our habitats for the changing tools and devices that will soon be entering the market—and our lives.

THE RISE OF MEDUSA . . .

You are having a great day, productive yet carefree. You open your closet door to put something away, when out of the corner of your eye, you glimpse what looks like a nest of interwoven snakes with flecks of silver flashing with the shifting light. Your heart begins to race and you contemplate, *fight or flight*? Should you shut the closet door as quickly as possible, or investigate further? You decide to move closer, inching slowly so that you can ward off the threat at any moment, and suddenly, you stop in your tracks as your thoughts shift from fluttery panic to frustrated anger. The snakes morph before your eyes into Medusa's head, a ball of abandoned computer cords that threaten to turn you into stone if you keep looking at them. But instead, you just keep staring with your mouth agape, because the snakes are mesmerizing. And they have names: USB, firewire, pin-dot, two-pin. And the snakes connect to so many obsolete devices—your old digital camera, that practically antique Palm Pilot, your grandma's dot-matrix printer, your shattered iPhone 3. This is not Medusa's head, as you originally thought. This is a monster you created.

An Intervention for Digital Hoarders: You've GOT to Get That Tech out of Your Life

In my house, the Medusa head used to be stashed in one lonely box in my closet. But over the past couple of years, the head has multiplied, filling my desk drawer, my upstairs cabinet, my attic, and even my garage. I try to pretend the heads don't exist, but every time I stumble upon them, I am grossed out. And overwhelmed.

How many of you still own a 300-pound TV? Maybe you are thinking, *I'd better keep it because maybe I'll need this big-ass TV again someday*? Or perhaps more charitably you rationalize, *Maybe somebody who can't afford a TV would really like this gargantuan piece of plastic that takes up half a living room?* It's time for a reality check: donation centers like the Salvation Army won't even accept televisions that were built before 2006.

Our nostalgia makes us into completely irrational technology hoarders. In the olden days, expensive products were designed to last a lifetime. But with the speed of technology now, cameras, phones, and laptops are practically obsolete within five years. We struggle with letting go of objects for a variety of reasons—they might be useful, you might need them, they have sentimental value, or they are worth a lot of money. We are suffering from what I call the **GOT Syndrome (Guilt Over Things Syndrome**). We feel like we have *got* to keep these things, even if we haven't touched them in weeks, months, years—or even decades (cassette tapes, anyone?).

We struggle with letting go of objects for a variety of reasons—they might be useful, you might need them, they have sentimental value, or they are worth a lot of money. We are suffering from what I call the **GOT Syndrome (Guilt Over Things Syndrome).**

In fact, researchers at Yale recently found that it can literally hurt your brain to come to terms with letting go. They identified that two areas in your brain connected with physical pain, the anterior cingulate cortex and the insula, light up in response to letting go of items you own and to which you feel a connection.[5] These are the same areas of the brain that light up when you feel physical pain from a paper cut or drinking coffee that's too hot. When it comes to physical things, merely touching an item can cause you to become more emotionally attached to it. In one study, researchers gave participants coffee mugs to touch and examine prior to participating in an auction.[6] The researchers varied the amount of time the participants were able to handle the mugs to see whether the time spent would have an effect on the amount of money participants would be willing to spend on the mugs during the auction. Participants who held the mugs longer were willing to pay over 60 percent more for the mugs than participants who held the mugs for shorter periods. The study concluded that the longer you touch an object, the greater the value you assign to it.

Tech adds two new complications to the GOT Syndrome. First, it's hard to let go of a gadget that we spent $600 on just a year ago, only to find that its value plummeted as soon as the next model came out. Second, these gadgets carry extremely sensitive data and we are not sure how to safely dispose of them. And so these Medusa heads made up of gadgets and gizmos pile up in dark corners and closets, their very existence in our homes proving that the stone-turning effect really works—we are paralyzed by our fear of letting go.

The Elusive Golden Snitch

Even if we are able to part with some of the digital chaos in our homes, that doesn't stop us from amassing further collections. The allure of the new tech promising greater productivity, efficiency, and happiness is hard to resist. Just like Harry Potter, who played the role of Seeker in the Hogwarts Quidditch matches, we find ourselves in an epic battle chasing down shiny golden orbs of new tech that seem to race ahead every time we get close to capturing them. We see the value in keeping up with new tech, but there seems to be no end to the game in sight. We think, *Perhaps this one gadget will help me never have to clean my house again . . .*

A few years ago, my parents gave my brother a Roomba robot vacuum cleaner for Christmas (I think they were trying to make a statement). The Roomba website boasts the following:

> *The Roomba is your partner for a cleaner home . . . how often is there too much to do and not enough time to do it? Pretty often. But imagine having an extra hand to help you keep up with daily cleaning and get more done every day. Designed with you and your unique home in mind, we're here to help. You and iRobot. Better Together.*

As one user wrote in his Amazon review:

November 9, 2014 by ***bdenton*** ★★★★★

NEVER VACUUM AGAIN!!!

I love it, love it, love it. I don't have to spend time vacuuming floors every-day [sic]. I don't have to worry about under the bed or the couch. It all gets done for me. It is so easy! Just push the Clean button and away it goes.

I have to say, I was a little jealous. With two kids, two dogs, a cat, and a husband, *I* was the one who needed the automated helper—not my single brother who traveled all the time. For two years, I eye-lusted over the unopened Roomba box in my brother's closet (i.e., The Island of Misfit Toys), wondering if I could sneak it out and return it one day without his noticing. And then, one day, it just disappeared. He never used it. He gave it away to *his housekeeper* (oh, the irony), who had become a family friend.

Why didn't the Roomba solve my brother's housekeeping problems? The problem with the Roomba was not the tech itself—it was his making the time and space to use yet another gadget in his daily life. Tech advertising is so effective because it plays upon our greatest desires: to have more time, to be more productive, to find greater happiness. We may buy into it, but until we make room for that tech in our lives, it won't be contributing to our happiness or productivity—it will only be making our Tech Medusa stronger.

Tech advertising is so effective because it plays upon our greatest desires: to have more time, to be more productive, to find greater happiness. We may buy into it, but until we make room for that tech in our lives, it won't be contributing to our happiness or productivity.

WE ARE RUNNING OUT OF SPACE

The overconsumption of digital stuff has the same effect on your brain as physical clutter. Whether it be your closet, your office desk, or your inbox, excess things in your surroundings can have a negative impact on your ability to focus and to process information. In fact, a team of UCLA researchers recently observed thirty-two Los Angeles families and found that, particularly for the mothers, stress hormones spiked during the time they spent dealing with their belongings.[7] That's exactly what neuroscientists at Princeton University found when they looked at people's task performance in an organized vs. disorganized environment. The results of the study showed that clutter in your surroundings competes for your attention, resulting in decreased performance and increased stress.[8] Even the size of our online inboxes can become a trigger for our mood. Similar to what multitasking does to your brain, clutter overloads your senses, makes you feel stressed, and impairs your ability to think creatively.[9]

We crave structure in our lives, yet the rapid transformation of tech leaves us unsure of how to create that sense of order. We keep trying to create routines for new habits without actually prioritizing the habits in our lives.

We crave structure in our lives, yet the rapid transformation of tech leaves us unsure of how to create that sense of order. We keep trying to create routines for new habits without actually prioritizing the habits in our lives. For instance, I have downloaded more exercise apps than I care to admit without actually following through on any of them. I use tech as a crutch, hoping that if I find that one app, it will create a lifetime of change. It's sort of like wishfully thinking a crutch will make you start walking again rather than just help you stay upright.

Why does all this insight not lead to better habits? Like kids on a hot summer day, we sometimes run headlong into the spray of technology just for the fun of it. But moments later, we realize that our clothes are drenched and we have no idea what to do next. Taking a moment to contemplate our actions can save us a lot of time and trouble in the long run.

When to Fuse It and When to Lose It

Will we let tech continue to drive us at hyper-speed, filling us with guilt every time we look at the graveyard of outdated devices that we have no idea what to do with? Or will we control the stream of tech in our lives, grounding ourselves in the bigger picture and using tech intentionally to achieve our potential? We need a practical system for knowing when it's worth it to integrate new tech into our lives and when it's time to lose it (tech) before *you* lose it.

In business, we determine optimal strategies by using cost-benefit analyses. But in our personal lives, there simply isn't time or energy to run fancy formulas about whether Candy Crush is good for our mental health and productivity. Nor would we want to. Moreover, we know from behavioral economics that humans are not always perfectly rational (OK, not even close to perfect). In fact, Princeton professor and *New York Times* bestselling author Daniel Kahneman explains that humans have two different systems of thinking, one that is fast, intuitive, and emotional, and the other that is slower, deliberate, and more logical. By slowing down our thought processes, we begin to become more aware of our illogical cognitive biases and can then start to make better decisions. By extension, as we embark on the journey of decluttering the digital chaos in our lives, let's strive to slow down our thought processes about what technology we want "to fuse or lose" in our lives so that we can make space for greater happiness in the future.

THE "REALLY?!" RULE

To slow down our thought processes, we need an in-the-moment mantra to decide what to do with the tech clutter in our lives. My dirty little secret is that I've got *a lot* of clutter in my house. A few years ago, I was attempting to declutter my house before listing it for sale and I was really struggling with deciding what to get rid of. My brother happened to be at the house that day and he obnoxiously started to follow me around, saying, "Really, Amy? Do you really need that?!" He's lucky he didn't get strangled. Nevertheless, his question echoed in my head so often that it actually began to work.

So I now present to you the "Really?!" Rule: *Does this tech truly make me happier and/or more productive?* For instance, does having a laptop in class help me focus or does it distract me? Does leaving my cellphone on my desk make me more responsive or do I feel tempted to check every alert and notification? Does sleeping with my phone on the bed give me good insight about my sleep habits or cause me to go to bed later because I wind up reading myself to sleep on my phone? The beauty of the "Really?!" Rule is that it allows for a wide range of answers and it can vary from day to day.

In *The Magic of Tidying Up*, bestselling author and Japanese cleaning consultant Marie Kondo teaches readers about her revolutionary KonMari Method for decluttering: simply put your hands on everything you own and ask whether that item "sparks joy."[10] While I love this approach, I actually take this question one step further, because with the blending of Work/Home/School/Play in our lives, some items have utility beyond pure joy. For instance, there are times when I play a mindless game on my phone just because it makes me happy; times when I play because the mental break actually makes me more productive; and times when playing doesn't make me feel happy or productive. When I recognize that I'm starting to fall off the "Happiness Cliff" like Wile E. Coyote, it's time to limit that activity in my life or cut it out altogether—and just like the earlier example of my husband and dad playing online games together, sometimes I may need my friends and family to intervene to help me see when I've gone off the ledge.

A recent study found that individuals who limited their frequency of checking email to three times a day experienced significantly lower daily stress, which in turn predicted higher well-being on a diverse range of well-being outcomes.

Similarly, another study found that the more constantly one checks social media, the less positive one's mood is.[11] As one user explains, "Although Facebook de-stresses me, I feel like it's a waste of time for me to go on Facebook. When I see myself going like half an hour or more, I think that's just a distraction and it's not productive."[12] Having this kind of self-awareness and self-control is rare, but this level of insight is exactly what we need to truly create habitats of happiness in our lives.

HIGH-IMPACT ORGANIZING

Knowing that the "More than Moore strategy" is about to take effect, it's time that we start creating spaces and systems in our lives to handle this. Gestalt psychology is a branch of traditional psychology that tries to understand how we acquire and maintain meaningful perceptions in an apparently chaotic world. In other words, Gestalt psychologists understand the impact of digital disruption in our lives and work to apply an idea called the Law of Prägnanz to reduce stress. The Law of Prägnanz says that we tend to order our experience in a manner that is regular, orderly, symmetrical, and simple, and the best way to decrease stress is by removing stimulating objects from our visual field. Gestalt theory appeals to the Type A part of myself, the side that loves Excel spreadsheets; the side that wants my life to be orderly, even if I know it will just get messed up as soon as my kids come home from school; the side of myself that craves "get-my-life-in-order days."

The Law of Prägnanz says that we tend to order our experience in a manner that is regular, orderly, symmetrical, and simple, and the best way to decrease stress is by removing stimulating objects from our visual field.

I think this must be why I love photo-organizing programs like Shutterfly, because they give structure to my previously chaotic storehouse of photos, letting me search through my treasured moments by date, keywords, "faces," or "places." Yet even so, I have just swapped my physical memory hoarding for digital memory hoarding, albeit slightly more organized. The real underlying question is *why* I am keeping so much—is it a fear that my children are growing up and I have no way to bottle up these memories? Or that someday my third child will wonder why her sisters have more pictures taken of them? Or that perhaps someone else in my family sphere may want these photos in the future? Or that one of my children will want to run for president someday and I'll be scolded if I can't fill in the early years for the presidential library?

The bottom line is that we can't keep everything. It's not physically possible, and it wouldn't make us that much happier. It's worth noting that while clutter has been shown to hamper performance, it is actually *your perception* of clutter that matters. Everyone's tolerance for clutter is different. For some people like Steve Jobs, a bit of a mess can feel inspiring and productive.[13] These individuals might see a clean office as a dormant area, an indication that no thought or work is being undertaken. On the other hand, people like TreeHugger founder Graham Hill prefer absolute simplicity. Hill was known for trading in his million-dollar mansion for a 420-square-foot apartment that has only the bare essentials in the kitchen: twelve salad bowls and basic utensils. Identifying your optimal environment by knowing yourself better will help you determine how much you need to declutter to find your happy place.

As you set out to regain control of the closets and drawers in your home, keep in mind that our goal is *high-impact organizing*—we are looking for the biggest bang for the buck. The first rule of thumb with organizing is to get rid of unnecessary items. Deleting spam is a low-impact activity because, like the Walking Dead, spam keeps coming back day after day, minute after minute. Your email program will likely delete the contents of your spam folder after every thirty or ninety days, so you're wasting time going through each email. So, just as you scan your inbox for emails from actual people and not mass-newsletters to read first, you need to start by identifying the biggest or most valuable items in your physical environment that you need to deal with and working out from there.

Physical Clutter: Performing a Sort

Although the process can seem daunting, devoting a day or even a weekend to decluttering can make a serious dent in your tech-device graveyard. Divide your unused gadgets into four piles: sell, donate, recycle, or keep.

Happy Hacks to Get You Started

1. **Sift and sort.** Start by sorting your items into two piles on the floor: those that are still used/needed and those that are not really truly needed. My general rule of thumb is that if I haven't used something in the last two years, I probably don't need it. Yes, that means parting with your old dial-up modem (trust me, it's never coming back in style).
2. **Identify redundancies.** Go through your "needed" pile to see if you have multiple copies of the same device or plug. Yes, one Ethernet cord can be handy in a pinch, but fifteen cords is overkill.
3. **Contain yourself.** To limit entropy in the future, put your "needed items" in small storage spaces and containers. Let me say that last part again: *small* storage spaces and containers. As we know from the Laws of Thermodynamics, particles expand to fill their spaces, so it would make sense that limiting our spaces would help contain the disorder. Avoid cardboard boxes since they are food for potential roaches and other bugs. Use clear (and lidded, if you can find them) containers so you can easily find your Medusa box, even if you don't have a label maker to mark it as such.
4. **Deal with the leftovers.** Divide your pile of "not-needed items" into three boxes: sell, donate, or recycle. For items that you are interested in selling, one of the fastest solutions I have found is taking my items to Best Buy, which offers to purchase an extensive list of items at a reasonable price. You can even use the company's online trade-in calculator to get a sense of how much you might get in return. If the item is not easily sold, you can also donate or recycle it. Try not to throw away electronics, if possible. According to the United Nations, the world threw away about 46 million tons of electronics in 2014, and the US alone was responsible for 7.7 million of those tons.[14] To find an organization in your area that accepts technology, check out e-Stewards.org, a nonprofit website with some

great information about how to donate your items. For example, old cellphones that you think no one may want are turned into lifelines for women fleeing domestic abuse. Note: Before parting with your items, make sure to remove sensitive information by securely erasing data. (If you are unsure of how to do this, look up "erase" and your particular device on Google. Or just find your nearest teenager to help. If all else fails, many cities have on-site "hard drive destruction" services, or you could even mail your devices to be destroyed through services like Ship'N'Shred.)[15]

5. **Own your space.** Straighten up your workspace so you have a visual reminder of all the work you just did.[16] Condense cords with cable ties or Velcro to eliminate the chance of "getting your wires crossed" and to increase the order in your field of view. This step may seem superfluous when you're thinking about how much other work you have to do, but it's an important way to keep yourself organized down the road and avoid the spaghetti mess that cords seem to create on their own with just a little bit of wiggle room.
6. **Relish your victory.** Last, take a deep breath and enjoy the space!

In the course of writing this book, I decided to put the idea of "putting my tech graveyard to rest" into action and learned some valuable lessons along the way. First, you probably will need to cut yourself some slack. I had visions of recovering entire closets of space over the weekend, but that was far from what happened. I have *a lot* of junk to sort through. If you're in the same boat, realize that it might take you longer than you expect. Marie Kondo advises that her technique can take at least six months to properly complete. You can make organizing and decluttering fun by recognizing the action you've taken. That's what I did. Using the Zorro Circle that we talked about in Strategy #3, I tackled one small drawer or shelf first and then defended it vigorously from little attackers who approached with crayon remnants and empty applesauce packets. This is my space! Once my brain perceived a victory, I gained renewed energy for tackling bigger projects, one Zorro Circle at a time. When I started to feel overwhelmed again, I repeated my mantra that I was "choosing happiness instead of defeat" and celebrated my progress to date.

My second lesson was that learning how to erase data on different types of broken devices was the most time-consuming part of this process. If you have a friend or know of a service that can speed up this process by digitizing your photos, videos, or slides, by all means use them! And finally, I recognized that creating a system for sorting was as important as actually parting with items. I might not be ready to part with that old boulder-sized camcorder, but at least I have restarted my project to digitize my videos, and after I finish those last few videos, I will be happy to part with it. Given the rapid rate of turnover of technology in our lives, this process of sifting through our storehouses of devices will be an ongoing part of our lives; starting on it now will make the future so much easier.

Computer Clutter: Creating Space

Next it's time to tackle the chaos on your computer. Too much clutter on your computer slows your entire system down, so one of the fastest strategies for boosting your productivity is to free up space on your computer. Rather than get bogged down by lots of little files, focus on high-impact items (movies, applications, photos, music) that are taking up a lot of computer memory.

Happy Hacks to Get You Started

1. **Use the cloud.** Save yourself the anxiety of wondering if your files will get lost due to a computer crash, lightning strike, or unfortunate tumble down the stairs (of which I have experienced all three now, *sigh*). Use an online service that backs up your computer automatically so that you don't even have to think about it. I actually back it up doubly now because some services are good for an entire system restore and others are better for accessing individual files. I use Carbonite for my primary backup and Dropbox for my secondary backup. I also store photos on Shutterfly and

videos on Vimeo. While this might sound like overkill to you, the benefit is that I know my files are safe *and* I know that I can easily access them (no more searching through multiple external hard drives or computers to find that one beloved photo!). Having this process in place means I can delete files off my computer that are slowing it down, which in turn means I can get more done with less frustration.

2. **Get rid of big files.** I like to sort items on my computer by size and then delete as many items over 500 MB that I can. If I am not ready to delete an item, then I try to transfer it to some sort of cloud storage (like iCloud or Dropbox).
3. **Identify redundancies.** There are a number of free and paid applications that you can run on your computer to help you delete multiple copies of the same files (like Mac CleanSweeper). The beauty of these apps is that you can start the programs and walk away while your computer does all the work for you!
4. **Archive old emails.** If you haven't read or looked at an email for more than a month or two, what are the chances that you are really going to go back and deal with it in your future free time? Really?! I like to archive emails older than one month, so that I can really see and focus on current and urgent emails in the present. If I do get inspired to get back to those emails, I know that they are still on my computer and I can always go back to them.
5. **Clean off the desktop for visual relief.** Lastly, move everything on your desktop screen into your Documents folder—or, better yet, into organized folders on a cloud-based storage system like Dropbox. A cluttered desktop is not only distracting but also makes it harder to think. Do yourself a favor and get a fresh start!

Notification Clutter: Hack the Distractions

Just as decluttering our spaces improves our productivity, efficiency, and creativity, so decluttering our minds serves to improve our mental functioning and overall mood. However, distractions in our environment are like weeds. Unless you remove the roots, they will continue coming back, causing you stress and anxiety. Finding ways to stem the streams of alerts and information will leave you more time and space to live a more creative and productive life, with less stress to boot![17,18]

Happy Hacks to Get You Started

1. **Unsubscribe from the unnecessary.** You can use a service like Unroll.me or just make a habit of clicking "unsubscribe me" at the bottom of promotional emails as they come into your inbox.
2. **Automate the mundane.** Save time and energy by letting online services help you tackle repetitive tasks like paying bills or backing up your computer.
3. **Turn off alerts.** Unnecessary notifications break your focus, slow your roll, and ruffle your feathers. Take five minutes to turn off as many notification alerts as you can on your phone or computer and save yourself hours of frustration and hassle later.
4. **Prune your acquaintances.** Cut back to only the friends you really want on social media (you can still stay friends with those other folks, but they won't show in your feeds).

As you walk through this process of decluttering your physical and digital space, I want to emphasize that the goal is not to arrive at some static state of perfect calm, cleanliness, and organization. Entropy just happens, as my children teach me daily. And entropy will continue to happen—you *will* amass more stuff, you *will* download new apps, you *will* receive more email, not to mention junk mail. But in the chaos of life, you have a choice about whether to be happy and it's a choice you have to practice, even in the midst of chaos and clutter. The point of the "Really?!" Rule is to help you feel empowered to reach your potential each step along the way, to give you more space to breathe and live in the now, and to create a more helpful framework and new thought process for dealing with new technology.

DESIGNING PLACES FOR GREATER HAPPINESS

Thanks to technology, we've become globalized, transient beings unbounded by the usual constraints of place and space. Some people work in offices, while others work from home, co-working spaces, or even coffee shops. We move fluidly in and out of a nebulous internet-based "cloud," with half of our belongings in the physical world and the other half in some virtual world. As the places that we live, work, and learn become increasingly blurred, we have to reconceptualize what happiness looks like in the modern era.

In *The How of Happiness,* renowned positive psychologist Sonja Lyubomirsky revealed that only 10 percent of our happiness is due to external circumstances, while the other 90 percent of our happiness is due to our perception of the world.[19] This groundbreaking research means that happiness is a personal choice, and that choice has powerful implications for our productivity and success. However, we are realizing now that this choice is one that we can influence for the better to encourage positive behaviors. As we explore ways that we can design places for greater happiness in the future, it's worth thinking about how we can use our environment to facilitate growth and a greater sense of happiness.

In college, my husband helped conduct a fascinating research study of the effects of environmental enrichment on individuals with Huntington's disease. Mice genetically engineered to mimic the symptoms of Huntington's disease were tracked over twelve weeks. Half of the mice were given an enriched environment (mazes, activities, stimulation), while the other half weren't. As predicted in previous research, the enriched mice ran faster and acted smarter; but my husband's research team took the study a step further by actually cutting open the brains of the mice whose environments had been enriched (gross, I know). In doing so, they learned that the brains of the enriched mice had actually changed *on the cellular level*: they had fewer pathologic protein deposits in the brain. Enriching the environment actually slowed the development of Huntington's disease, a groundbreaking finding that revealed the immense power of environments in shaping the brain—and behavior.[20,21] What I love about this study is that simply changing the environment for twelve weeks creates a structural change in our brains for better performance! In the coming pages, I want to share with you some practical strategies for how you can enrich the places that you live, work, and learn to achieve greater happiness and productivity.

WHERE WE LIVE

When I was in middle school, my parents were both working full-time. So after school, my brother and I would walk home and let ourselves into the house. Plopping down our school bags, we would grab a snack, find a seat on the couch, and, of course, turn on the TV. All of this would have to happen quickly, though, because at 4 PM, *The Jetsons* would come on (and there was no TiVo back then, gasp!). We knew every episode by heart, but that didn't stop us from watching them again and again . . . that is, until we heard the garage door opening. With military precision, we would jump off the couch and spring into action, turning off the TV, straightening the couch pillows, even swiping here and there with the duster to make it look as if we had done some chores. But invariably my dad would walk in and see the afterimage of *The Jetsons* on the old analog television set, and we were busted.

If only I had the words back then to explain to my dad that I was just doing research for a book that I would one day write about the future (I can't wait to put a copy of this book in his hands for proof). What I didn't know back then was that today's political, social, and business leaders were also watching *The Jetsons* at home on repeat, shaping a collective vision for the future.

The original *Jetsons* series aired in 1963, shortly after the end of the Cold War and the founding of NASA in 1958. Americans seemed equally optimistic and terrified about the future, giving rise to what became called the Golden Age of Futurism. It was a time filled with artistic renderings of techno-utopian dreams of life in the future.[22] Danny Graydon, the London-based author of the official guide to *The Jetsons*, explains: "It [*The Jetsons*] coincided with this period of American history when there was a renewed hope—the beginning of the '60s, sort of pre-Vietnam, when Kennedy was in power. So there was something very attractive about the nuclear family with good honest values thriving well into the future. I think that chimed with the zeitgeist of the American culture of the time." Two decades later, *The Jetsons* was rebooted in color, exposing a whole new generation (including me) to its visions of the not-so-distant future.

Other artists were captivated by the idea of futurism as well. In 1957, artist Arthur Radebaugh created a syndicated cartoon called *Closer Than We Think* that featured jetpacks, meal pills, flying cars, and more.[23] That same year, Disneyland introduced the Monsanto House of the Future, a stand-alone building with a curved spaceship-style exterior made entirely of plastic. Outfitted with modern furniture and space-age appliances like the microwave, the futuristic house was predicted by the designers to become reality before the millennium. Amazingly, *The Jetsons*, Radebaugh, and Disney accurately portrayed technologies common to all of us now, including flat-screen TVs, video calls, interactive news, talking alarm clocks, and even watches that can stream content.[24]

It's hard to know whether these artistic renderings of the future created a self-fulfilling prophecy or whether they were simply ahead of their time. We may never know the answer to this chicken-or-egg question, but what we do know is that the technological revolution is changing the places where we live in some fundamental ways.

Our Homes

I was shocked when I learned that Disneyland had torn down the Monsanto House of the Future last fall. Apparently our homes have caught up to the future, becoming "smarter" than ever before. With everything from lighting to curtains to security to pools being controlled with the click of a smartphone button, what could possibly come next? Brace yourself for hyper-speed, because Disney's latest version of the future is rumored to be a virtual-reality trip on board a spaceship: specifically, the Millennium Falcon.[25] While the ride is not intended to be an updated house of the future (it's actually part of a new *Star Wars*–themed Land), Disney continues to push the bounds of our imagination by offering a glimpse into an intergalactic life composed of humanoids and droids. We might love the short-term immersive experience, but would we really want the future to look like this? How soon would we morph from handsome Han Solos into the walking-challenged spaceship inhabitants of *WALL-E*?

Interestingly, both of my parents echoed similar sentiments when they toured the House of Monsanto for the first time in the 1950s. At the time, my parents were teenagers living in Anaheim, California, and working at Disney. My mom was a ride operator and my dad was an ice cream man who enjoyed making sure that no princess walked by without getting a wink and an ice cream. When I asked them about their impressions of the Monsanto House, both mentioned that they had visited only once or twice out of the hundreds of times they'd gone to the park. They explained that the Monsanto House was fascinating, but everything seemed so sterile and sleek, lacking in color and vibrancy. My parents, like many tourists, just wanted a glimpse of the future, but they had no interest in actually living there.

Yet here we are living in homes outfitted with those newfangled contraptions that the Monsanto House featured called *microwaves*—and most of us can't imagine not having them in our lives today. Soon our homes will be brimming with the next iteration of the previously unimaginable. I had the opportunity to tour a "kitchen of the future" at the 2015 World Expo in Milan, which featured jaw-dropping technology like a fridge that uses your fingerprint to recommend what you should be eating based on personalized health data and a pantry that tells you which groceries you are running low on so that you can stay on your diet. I even tried on Oculus goggles to walk through a virtual world and learn from which farm or ocean each item of food in the kitchen had come.

Our homes of the future will seamlessly integrate our digital and physical worlds, linking everything together over Wi-Fi to become part of this crazy network that drives our lives called the "Internet of Things." The idea behind this shift is that by automating and aggregating tasks and information in our environment, we can increase safety, save energy, and reduce anxiety. For instance, Apple's new Home app boasts the ability for users to "control a Fantasia-like orchestra of smart gadgets from one place, including everything from smart doorbells and locks, to thermostats, light bulbs, humidifiers and entertainment systems."[26] Did you forget to lock your door when you left home? Don't worry, your smart home profile will lock the door when you leave, turn off extra lights, and perhaps even adjust your thermostat to save on energy costs. In fact, you can even answer your door with the Ring Video doorbell that lets you see and talk remotely when someone comes to your door.

Certainly, smart homes sound nice and helpful, but at the end of the day what I really want is a smart *and happy* home. Having more devices, more cords, and more apps isn't the solution. Rather, I want to consolidate my devices, and I want the devices that I do have to make my life better and to run seamlessly in the background so that I don't even have to think about them.

> The happiest places are ones that are connected not with physical wires but with deep personal and emotional circuitry.

The happiest places are ones that are connected not with physical wires but with deep personal and emotional circuitry. So while I agree that there is certainly a time and a place to unplug to reconnect, there are ways that we can use existing technology in our homes to raise engagement levels and happiness. I encourage you to avoid the road of the tech doomsday-sayers, because I don't see that it is truly possible for us to eliminate technology and I don't think we should have to eliminate technology to find happiness. Instead, we can rethink the ways that we engage with each other around technology. For instance, if your family tends to congregate around the television, set your screensaver to cycle through family photos as a form of communal journal, thus creating an opportunity to relive the most meaningful experiences of life together. Or if everyone in your family is already using social media, leverage Instagram or Facebook to share family gratitudes to collectively train your brain for greater optimism and use a custom hashtag to track your submissions over time (#happyblanksons). Designing your home for greater happiness doesn't have to be complicated or expensive; it just means being thoughtful and intentional about what brings the greatest level of meaning to your environment.

Our Communities

As influential as physical spaces can be in our lives, a culture of connection within our spaces is even more powerful. And no one understands that better than Habitat for Humanity. Habitat for Humanity has a vision for the future: that no one should have to go without a place to live. Since its founding by President Jimmy Carter in 1976, Habitat for Humanity has partnered with volunteers to build more than 800,000 homes around the world. When we were filming an *Oprah* e-course on happiness, I had the opportunity to talk with Jonathan Reckford, CEO of Habitat for Humanity, about what makes his organization so successful, and he explained that the organization's focus is not just on creating homes, but on creating *habitats* for happiness. If, for instance, a beaver were to try and build its dam on a busy highway, it would be disastrous. Instead, a beaver needs an entire ecosystem that is conducive to thriving, including a safe stream, ample wood, and a food supply. In the same way, Habitat seeks to create that ecosystem for families in need, as well as for the whole community where the new home is being built. Future homeowners contribute 400 hours of "sweat equity" toward building their own home plus the building of other families' homes in the neighborhood. Also, volunteers in the community work on homes side by side with the families in need. Habitat is even intentional about connecting office staff back to the mission. As staff members leave each day, there are signs by the elevators saying, "Thanks to your efforts, 481 additional people have a place to sleep tonight."

While Habitat for Humanity's model is incredibly effective at building communities for future happiness, we also have an opportunity to create a culture of connection right in our own neighborhoods. An emerging new field called civic technology aims to do just that, by leveraging digital tools to serve the public good.[27] If we truly want to shape the future of the spaces we live in—and not just be tourists glimpsing it from a distance—now is the time to embrace the spirit of opportunity by being intentional about creating meaningful community. I will talk more in the next section about how you can shape the places where we live by being a conscious innovator of positive change.

WHERE WE WORK

The Google Effect

Free sushi, a pampering massage, and an extended catnap in the sunshine don't guarantee your happiness, but these things certainly help.

That is the philosophy of Google, whose primary goal is "to create the happiest, most productive workplace in the world."

In 2005, Google executive Larry Page reached out to architect Clive Wilkinson to discuss a redesign of the Google campus in Silicon Valley. Wilkinson maintains the belief that "[c]ubicles are the worst—like chicken farming. They are humiliating, disenfranchising, and isolating." Wilkerson and his team persuaded Google to move away from the traditional cubicle model of office design toward a more flexible and innovative workspace that aligned with their overall brand, and the rest is history. Google is now renowned for its playful workspaces, where there are bumper-car desks and red telephone-video booths; conferencing stations inside giant, primary-colored dice; tube slides between floors; and more—and when I say more, I mean massage rooms, grand pianos, and complimentary services like eyebrow threading and yoga sessions.[28] Oh, and yes, all the sushi you can eat.

In 2014, I visited the Mountain View campus for the first time while working on a consulting project. I already knew what to expect, but that didn't stop me from drooling as I walked past the numerous micro-kitchens, strategically located throughout the campus so that I was never more than 100 feet from fresh, healthful snacks. After lunch, I hopped on to one of the free Google bikes to cruise around campus and see the swimming pool, packed with employees enjoying a friendly game of Marco Polo. No wonder Google employees are happy!

For those who have worked in more traditional offices, you may skeptically wonder how much work actually gets done in that environment. Yet Lauren Geremia, who has designed office spaces for the likes of Google, Facebook, Dropbox, and Lumosity, explains that these companies realize that "the talent is so indispensable that they've linked the success of the company to how happy people are in the workplace." While the Google office "experience" has been compared to that of a college dorm, beyond the cool furniture and jaw-dropping amenities is an organizational philosophy that work can and should be a place that inspires and motivates. *Fun buildings are great, but what is inside is more important.*

Laszlo Bock, senior vice president of people operations at Google, knows that the initial wow-effect at Google is not enough to sustain the employee engagement for the long run. He says, "It's no surprise that Nooglers (new Google employees) start full of optimism and excitement, but on average that ebullience trends ever-so-slightly downward over time. I often jokingly tell them to brace themselves, because they will never be happier here than on their first day. It's all downhill from there! You've probably felt this in your own jobs: Day one is awesome. Day one-thousand-and-one, not so much."[29] Bock is describing what is known in psychology as the hedonic treadmill, or the tendency to return to a baseline level of happiness despite major positive or negative events or big life changes.

Blogger Sidd Finch parodied this experience:

> *So you take the fancy job and have your first week at Schmoogle. It's freaking awesome. It feels like you're back in college! You spend lunch time in the cafe meeting new friends, ride bikes during the break time, and get tons of new supplies like a MacBook Air, a fancy monitor, and a new phone. "God, this is awesome," you think. Then, in your second week, work starts. You knew it was coming, but the first week was so much fun that you totally forgot about it. And, as it just so happens, during your first week of real work the cafe at Schmoogle runs out of Cinnamon Toast Crunch, your favorite cereal. When you ask the kitchen staff if they plan on restocking soon, they say that because of the CEO's new gluten allergy they'll no longer be having Cinnamon Toast Crunch. It's not like you started working at Schmoogle because of the free cereal, but it certainly didn't hurt. Again, not a big deal, but Schmoogle just went down a little bit in your book.*[30]

Curious about the sustainability of the "wow-effect," in 2014 I embarked on a research project with Alex Chanson, account specialist at Google, who was tasked with onboarding Nooglers. Alex explained to me that many Nooglers "tend to think that if they get a job at Google, they'll be *permanently happy,*" but this initial boost fades quickly as reality sets in. To actively reduce this effect, Alex redesigned the Noogler orientation program to include principles of positive psychology like gratitude and mindfulness. Together, we then designed a survey to measure the happiness of Nooglers over their first month of training, taking measurements each week to track the differences in attitudes and outlook. At the end of the four weeks, there was actually an 8 percent *increase* in the number of Nooglers who felt that Google was a great place to work. Chanson explained that the training seemed to help Nooglers shift from the thought process of "How can Google make me happy?" to more of "What can I do to increase well-being at Google?"

This hypothesis was confirmed by a new research study that Laszlo Bock quoted in his recent *Harvard Business Review* article, revealing that "employees who self-identify as more grateful are largely immune to the sinking effects of tenure on satisfaction. They stay happier, longer."[31] Recently, Google's People Innovation Lab took this idea one step further, launching what it hopes will be a century-long study aimed at understanding how happiness affects work and how work affects happiness.[32]

Google understands that, while environment can influence our happiness, it does not determine it.

Google understands that, while environment can influence our happiness, it does not determine it. A good friend of mine once worked for a large tech firm that tried to re-create the Google vibe by installing a slide to get from floor to floor and putting ping-pong tables in open spaces. However, my friend describes his experience at that company as one of the most toxic environments he has ever worked in because of the disconnect between managers and employees. Which would you rather have: a positive, engaged workforce in a crummy building, or a crummy workforce in an awesome building? (That's sort of a trick question, because if we truly had the choice we'd all rather have an awesome workforce in an awesome building!)

Individuals make a company culture. Whether you are a manager or a front-line employee, each of us has a choice to shape the environment in which we work through our mindset and our actions. We may not always have a choice about our work schedules or our work priorities; however, there are powerful things that we can do to regain a sense of control about our happiness at work.

Individuals make a company culture. Whether you are a manager or a front-line employee, each of us has a choice to shape the environment in which we work through our mindset and our actions.

Happy Hacks to Get You Started

1. **Strategically unplug.** In the Digital Era, most employers expect employees to be plugged in via email, phone, text, instant message, or all of the above. This constant barrage of communication can be incredibly frustrating, if not counterproductive. While completely unplugging from email or phone may not be realistic in today's working world, stepping away from technology, even briefly, can increase your focus, which leads to a 57 percent increase in more effective collaboration, an 88 percent increase in learning effectiveness, and a 42 percent increase in socializing effectiveness.[33]

Stepping away from technology, even briefly, can increase your focus, which leads to a 57 percent increase in more effective collaboration, an 88 percent increase in learning effectiveness, and a 42 percent increase in socializing effectiveness.

If you have the flexibility and permission from your boss, consider setting a short-term auto-responder directed specifically to others in your company explaining what you are doing and when you will be back (i.e., I'm stepping away from my email to finish this project. I'll be back in one hour). I mentioned earlier that Shawn used this approach at our company when he was trying to meet a project deadline or take a vacation, and the gesture gave the team such peace of mind because we knew exactly how to shift our expectations for a response. You may be surprised how many employers are actually thrilled that you want to focus more (and even inspired by your initiative to communicate this because they secretly want to do the same thing).

2. **Hide your phone.** In addition to taking breaks from email, you also can hide your cellphone. Chances are, seeing every text message, email, or social media alert as it comes in won't make you happier, and it certainly won't make you more productive. Recent research shows that *the mere presence* of a cellphone can decrease your productivity and attention on cognitively demanding tasks.[34] To focus on your work, move your cellphone out of your line of sight (put it in your bag, behind your computer screen, or in a drawer); if that's not possible, at the very least, turn off nonessential notifications. You also can get noise-canceling headphones to help you focus.

Recent research shows that *the mere presence* of a cellphone can decrease your productivity and attention on cognitively demanding tasks.

3. **Infuse your workspace with meaning.** Many individuals make the mistake of just trying to slog through work until they can get home to their "real life." However, research is revealing that investing in your personal space at work is beneficial to your happiness and health, and it can also increase your productivity by 15 percent.[35] Amy Wrzesniewski, an associate professor of organizational behavior at Yale University's School of Management, found that individuals who connect to their work on a deeper level are more satisfied in general with their work and their lives. Whether you have a cubicle or a shared space, take the time to find little ways to connect your home life with your work life, whether through photos and art on the walls or something more transportable like a day planner. For an extra boost, bring a plant to work to lower stress, reduce blood pressure, and feel more attentive.[36]

Research is revealing that investing in your personal space at work is beneficial to your happiness and health, and it can also increase your productivity by 15 percent.

4. **Breathe life into your workspace.** One of the best ways to shape how you feel about your work environment is to start the day by meditating. Meditating means different things to different people, but for the purpose of this exercise, here is what I suggest: When you arrive at the office, take two minutes before you ever turn on your computer just to close your eyes and focus on breathing in and out. Carving out a few minutes for silence and concentrated breathing can not only transform the way that you see your work, but it also improves accuracy (by 10 percent, according to some tests), boosts your happiness, and decreases the stress level of others on your team—even if they are not meditating.[37] This quiet time at the beginning of your day calms your mind down enough to start work with intentionality rather than with a which-fire-should-I-put-out-first mentality.

   You also can do this when you return to your desk from lunch (it is good to actually take a break for lunch rather than eat at one's desk day after day, even if it's just a little walk around the building to get a happiness boost from being outdoors, away from tech *and* away from work). Breaking the workplace culture of eating lunch at your desk and staying late to skip happy hour and finish a project is not easy! We have to consciously unlearn the habits we were taught by our parents, professors, and bosses. Even if the culture at your workplace is still using the nose-to-the-grindstone approach, you have to take the brave step to make these minor (and legally allowed!) changes to actually take your fifteen-minute breaks, eat lunch outside or at least away from your desk, and fight the fires only after you've taken two minutes to breathe each morning.

5. **Create a culture of connection.** Speaking of work lunch habits, you would not believe the number of people I meet who say that they eat lunch alone every day. In *The Happiness Advantage*, Shawn writes that social support is one of the three strongest predictors of your long-term happiness. One of my favorite statistics is that social support is just as predictive of how long you will live as obesity, high blood pressure, and smoking. Wow! Some people explain that there is no common space for eating or that they eat lunch on the run so they can get home sooner. However, these same individuals are the ones who express dissatisfaction about lack of connection in the office.

Social support is just as predictive of how long you will live as obesity, high blood pressure, and smoking.

Knowing the importance of an engaged workforce, many employers have started to include questions on their annual surveys about how much social support employees are receiving. However, at GoodThink, we decided to study what would happen if you flipped this question around. Instead of asking how much support employees were receiving, we began to ask employees questions like: Do you help others when they are behind on their work? Do you initiate social support? Do people see you as someone they can talk to? And after administering this survey to thousands of employees, we found that employees who *provided* social support to others were ten times more engaged at work and 40 percent more likely to receive support in return.

If you want to create a connected culture, it's up to you to create it. Start a post-work happy hour, invite a co-worker to lunch, learn the names of all of the employees in your vicinity, make a practice of saying hello to others in the building. These simple practices are the foundation of creating a connected culture and ultimately shaping a social script for workplace happiness in the future. And remember, anyone, including the lowest person on the totem pole, can practice these habits.

If You Are a Manager . . .

While employees are ultimately responsible for their own happiness at work, managers can influence the company culture by fostering an environment that maximizes employee happiness. For years, companies used a hierarchical system to assign offices—if you landed a corner office with windows, you would know that you had arrived. Yet workers today are questioning: what good is a fancy office with floor-to-ceiling windows if you never have time to look out the windows and benefit from the beautiful vista designed to increase your creative thinking?

Architect Clive Wilkinson believes that more American companies need to catch on to the way that firms like Google design their workspaces. A survey by *Management Today* found that 94 percent of respondents regarded their place of work as a symbol of whether or not they were valued by their employer, but only 39 percent thought that their office had been designed with people in mind. And 25 percent of American office workers actually found their workplaces to be downright gloomy or depressing.[38] Wilkinson continues, "The way we work is changing, but too often our habits are not. If we want people to work differently, then we need to help them make the transition."[39]

Ben Waber, the founder of Sociometric Solutions in Boston, explains that "the biggest driver of performance in complex industries like software is serendipitous interaction. For this to happen, you also need to shape a community. That means if you're stressed, there's someone to help, to take up the slack. If friends surround you, you're happier, you're more loyal, and you're more productive. Google looks at this holistically. It's the antithesis of the old factory model, where people were just cogs in a machine."

Although there's no universal agreement on what "good company culture" is, there's no denying that workplace design plays a role in influencing and shaping employee behavior. Not every company can or even wants to be Google, but for companies looking for the fastest interventions to create change in their culture, here are some of the top strategies for employee well-being that any company, of any size, can replicate.

Happy Hacks to Get You Started

1. **Encourage collisionable spaces.** Rearrange the furniture in your office to encourage the accounting department to walk by the legal department more. Place signs on various entrances and exits encouraging employees to use a main entrance. Simply shifting the office layout or desk setup can actually have a big impact on the flow of communication within the office by increasing serendipitous encounters. In a recent study, 60 percent of respondents to a survey conducted by the British Council for Offices said that having a nontraditional workspace improved their productivity.[40] Architect and designer David Rockwell told *Bloomberg Business* that he facilitates spontaneous interactions by creating "rivers of activity" in the office space. He suggests that companies should "[w]iden the hallways and put a reward at the end of them . . . if you want to encourage free movement in an office, have some kind of landmark. Traditionally it was a water cooler, or it could be a clock. But now it can be just a conference room."[41] Flexible meeting spaces also offer a space for spontaneous collisions. Office designer Lauren Geremia prides herself on creating more spaces for collaboration than necessary. Having more than one conference room or gathering area encourages employees to meet up and work together with ease.
2. **Invest in employee well-being.** This sounds like a no-brainer, but as my brother often says, "Common sense is not always common action." By providing healthful foods in the office, incentivizing employees to move around the office more, and offering comfortable seating, companies can actually

save money on health-care costs in the long run. Plus, in general, healthier workers are happier, and best of all, more productive.[42] Alex Haslam explains that "[n]ot only does office design determine whether people's backs ache, it has the potential to affect how much they accomplish, how much initiative they take, and their overall professional satisfaction . . . All this could have a huge impact for firms of any size, yet employers rarely consider the psychological ramifications of the way they manage space. By paying more attention to employees' needs they can boost well-being and productivity at minimal cost."[43]

With that in mind Cigna decided to take an innovative approach to well-being, offering a "two-minute staycation" in the extremely popular Cigna Relaxation Pod.[44] Employees are invited to sit in a comfy, egg-like chair, don a pair of Oculus virtual-reality goggles, and immerse themselves in a two-minute guided meditation to de-stress and calm their minds before returning to work. If your budget doesn't allow for this, think about some other, lower-cost DIY initiatives that you could do in your office (and share with me on social media so that I can help spread the word on HappyTechBlog.com).

3. **Create innovative spaces.** Simply put, if you want employees to be more innovative, their work spaces need to be innovative. This might mean replacing a desk and chair with a few colorful floor pillows, or switching a dim ceiling light with a bright funky lamp. Or you can follow Deloitte's "hot-desking" model, where desks are only used when they are needed. Instead, Deloitte has 1,000 desks for 2,500 employees and offers comfortable lounge chairs or coffee bars for other employees to use at will. This *het nieuwe werken* ("new way of working") philosophy is designed to break people away from their fixed locations and rigid ways of thinking, instead encouraging new relationships, chance interactions, and, just as important, efficient use of space.[45]

4. **Show off your employees' best ideas.** Visually highlighting an employee's work encourages other employees to strive for recognition. The MIT Media Lab Complex, designed by Fumihiko Maki, took this idea so far as to transform an otherwise industrial, tech-heavy space into a bright, museum-like public display where one could watch the future unfold. The goal was to create maximum transparency to allow for greater collaboration, competition, innovation, and democratization.[46] As Dimitris Papanikolaou, a recent Media Lab alumnus, describes, "The Media Lab is an enormous beehive of radical creativity and unconventional learning that brings together ideas, people, technology and resources to collaboratively materialize the craziest projects. With its reputation of breaking down disciplinary boundaries sometimes it feels like a science-fiction spaceship, other times like a messed-up kindergarten and other times like a pristine museum."[47]

WHERE WE LEARN

Imagine future schools in which students are totally engaged in a class. If you envision students sitting in a classroom, think again. "Next century, schools as we know them will no longer exist," says a feature in *The Age* publication, based in Melbourne, Australia. "In their place will be community-style centers operating seven days a week, twenty-four hours a day." If you are struggling to picture this transition, you are not alone. The Education Department of Australia actually decided to create an imaginary school called Seashore Primary School to model how this might work.[48] Seashore's acting principal explains that "[technology] has changed the emphasis to the learning of kids rather than the teaching of kids."

The US-based Fusion Academy founder Michelle Rose Gilman couldn't agree more. In 1989, Gilman was an innovative educator who was excited to change the world. However, when she began teaching, she quickly became frustrated with schools' being understaffed, underfunded, and uninspiring. She saw the emotional and social parts of students' lives being ignored, resulting in their stagnating rather than flourishing. So she began what was initially a tutoring and mentoring model and eventually grew into the Fusion Academy, "a revolutionary community of learning where positive relationships and one-to-one classrooms unlock the academic potential in every student and create opportunities for emotional and social growth." Fusion now has a network of accredited private middle and high schools for grades six to twelve that offers self-paced classes and tailored course material to fit every student's interests, strengths, challenges, and learning styles.

While the one-to-one Fusion model is intentionally designed to meet the needs of unique learning styles and not be a blanket solution, the idea of student-driven learning is catching on. Collaborations like the Global Digital Citizen Foundation are now collecting best practices and disseminating ideas to educators in different countries, helping them develop modern learning environments in their schools and classrooms that guide students toward taking ownership for their learning and developing critical thinking vital to their growth and independence.

Designing "habitats for learning" centers around flexible space for education and greater collaboration. In the same way that traditional office cubicles have transitioned toward co-working spaces, so rows of desks will become mobile classrooms with roving teachers. When I visited the XPRIZE Foundation, I remember seeing a remote workstation zipping through the office—it was literally a self-propelled computer on wheels controlled by someone on the other side of the world who was "visiting" the office to collaborate on a project. Just a few months later, I toured the new state-of-the-art building named Evans Hall at the Yale School of Management and saw similar capabilities—the only difference was that students were actually required to collaborate with other business-school students around the world on every project to learn how to navigate the globalized economy.

While the best piece of classroom technology is, and always will be, a human teacher, technology can certainly augment these powerful human connections. Remote learning enables students to take classes with some of the best teachers in the world. For instance, Yale University now offers "Yale Online" to the general public for free, providing access to world-class faculty on a range of topics from organic chemistry to game theory to modern poetry (visit: https://www.coursera.org/yale). Other universities are rapidly turning to this model as well, offering unprecedented access to a range of topics and teachers.

While the best piece of classroom technology is, and always will be, a human teacher, technology can certainly augment these powerful human connections. Remote learning enables students to take classes with some of the best teachers in the world.

Technology also facilitates learning from colleagues. In the military, an online medical subspecialty network was created to help isolated providers, like doctors stationed on remote islands, in submarines, or in foreign countries. Through the network, these isolated providers could send questions back to colleagues at major medical centers to get medical advice and timely aid to save lives.

Technology creates more opportunities for all sorts of students to learn and can greatly augment what a human teacher can do. In Bobo's military medical training program, doctors practiced emergency scenarios on incredibly realistic mannequins that had heartbeats, pulses, lung sounds, and even veins for intravenous access. A human instructor operating a wireless motherboard could change all of these features in real time to simulate different scenarios.

On the teaching side, research firms are now offering biometric feedback in real time to help educators or presenters hone their content and style. For instance, biometric feedback technology can tell us if the audience was engaged, if their eyes met the presenter, which stories led to the greatest surprise or joy, and what content left listeners feeling frustrated or confused.

To understand how this technology works, I met with Dane Hylen at iMotions Lab in Boston to try the software for myself. Using a simple computer camera, the iMotions software was able to track where my eyes were looking at all times and to "read" the emotions on my face down to the second. (If you want to try this software yourself, you can download the AffdexMe app for free and experiment with how the software can analyze what you are feeling using 40,000 different data points on your face. [Visit: http://bit.ly/affdexme])

The possibilities for this software range from corporate training and marketing to gaming to medical rehabilitation. My husband researched neurodegenerative diseases in college and in medical school, and he points out that this software could revolutionize how people with Parkinson's disease experience the world around them. Parkinson's disease is a movement disorder; the afflicted can sometimes struggle because their facial muscles prevent expression of emotions like happiness, sadness, humor, and even frustration, which in turn leads to feeling misunderstood and isolated. Imagine if people with Parkinson's could use this software to practice their joy, fear, and anger facial responses. Imagine the improved richness and depth of interaction they'd enjoy!

Lane Weiss, former superintendent of Saratoga Union School District, understands both the challenges and opportunities that technology brings to the classroom. Located in the heart of Silicon Valley, the public school serves the children of executives from Google, Facebook, Apple, and others. He sees children daily who are tethered to communication devices, and yet are struggling to communicate and express themselves in the classroom. I first met Lane in 2013, when he reached out to GoodThink after hearing Shawn speak at a conference. I remember thinking at the time, "This guy is the real deal!" As a positive-psychology enthusiast and devoted educator, Lane is genuinely one of the happiest people you will ever meet (after years in education, he still brings an enthusiasm to everything he does, including hosting a *weekly* barbecue for all district-office staff in the summers!). He has seemingly been involved in every major conversation at the intersection of technology, education, and positive psychology—from participating in original research with the University of British Columbia, to developing/piloting the MindUP curriculum with the Goldie Hawn Foundation, to designing an app to foster teen resiliency with the Dorothy Batten Foundation.

One of his favorite projects with the Goldie Hawn Foundation is called Taxi Dog, which is a television adaptation of the popular children's book featured on LeVar Burton's *Reading Rainbow*. The Taxi Dog program uses entertaining videos and interactive puppets to teach social emotional learning (self-awareness, self-management, social awareness, relationship skills, and responsible decision-making) to elementary school children while also fostering executive function skills like problem-solving, focus, and self-control.[49] The most significant part of the Taxi Dog program is the collaboration among tech, media, educators, and top-notch researchers that grounds the program in quantifiable metrics to prove that these skills can be taught effectively. As we seek to design places to be at the forefront of learning in the digital era, the solution is less about where and more about *how* we can work together to glean knowledge from every sector.

BUILDING INVISIBLE FENCES

Growing up, I spent most of my formative years in Texas before meeting my husband and marrying into the military life. As you may have heard, Texans love their land and are notorious for fiercely defending their property lines. If you were to take an aerial snapshot of a suburban neighborhood, you might actually confuse it with an advertisement for The Container Store. So you can imagine it was quite an adjustment when we moved to Virginia and found ourselves in a neighborhood with no fences. One yard sprawled into the next, with the only lines between houses created by intermittent lawn mowing.

At first, this setup felt strange; but soon, I came to love the openness and freedom that came with it. So did our dog. Within a few short weeks, our well-behaved dog began to wander farther and farther; she particularly loved to visit the other neighbor dogs. She soon discovered that most other dogs had invisible fences containing them in their lots, while she did not; so she would intentionally dash across the hidden boundaries to tease the other dogs and duck back into our yard just before the other dog caught her. At first, it was all fun and games; but we soon discovered that she was chasing the mailman, surprising unsuspecting mothers with their baby joggers, and wandering so far afield that she didn't know how to get home. Yes, I confess: we were *those* neighbors.

Fortunately, we learn through our mistakes. There's an old saying, "Good fences make good neighbors," so I promptly got a quote to install an invisible fence, and in the process, I learned something fascinating: most owners of invisible fences do not even have their systems turned on because once a dog learns where the boundary is, he or she will not cross the line again. While humans are a bit more complicated, we can use the same principle of the invisible fence to create healthy boundaries for using tech that will help us avoid temptation, keep us out of trouble, and lead to greater happiness in the long run.

DECIDING WHERE TO PUT THE FENCE

One of the biggest challenges in building an invisible fence is knowing where to build it. Recently, I spoke with one mother who had been feeling tremendous pressure from her peers to monitor her teenage son's text messages. Although he had never given her any reason to mistrust him, the mother's friends were adamant that a responsible parent should be monitoring her child's messages, lest he post explicit material or fall prey to internet predators or cyberbullying.

One night she was lying in bed reading on her iPad when an operating-system update suddenly made it possible for her to receive copies of all of her son's text messages on her device, as well. She secretly thought that it was providential intervention, relieving her of the moral dilemma about whether to monitor her son's texts. As ping after ping of text notifications came through, the mother succumbed to temptation and she began to read the string of messages. The mother's jaw dropped as her son confessed that he had a test the next day but hadn't studied yet (and he didn't even plan to!). The mother paused to ask her husband what she should do about this information, when her son walked into the room holding his iPhone. At that exact moment, both the iPad and iPhone beeped with an incoming text. Realizing what was happening, the son confronted his mother. Caught in the crossfire of peer pressure, motherly concern, and a crisis of conscience, the mother looked her son in the eyes and flat out denied her actions. The son's face visibly fell and he began to retreat from the room, but not fast enough that the mother did not catch the hurt and disappointment on his face.

In that moment, the mother realized that she had been the one to breach his trust. She went to his room and apologized immediately. The son acknowledged, "I couldn't believe that you lied to my face, but I'm really glad you came to talk to me." After a long conversation about boundaries and expectations, they hugged and then the mother said with a smile, "You *are* going to study for that test, aren't you?" The son laughed and said, "Mom, I *always* study, but there's no way that I'm going to let my friends know that!"

In the absence of a playbook for parenting in the Digital Era, we all stumble in trying to establish boundaries that convey love but also provide structure. Not all parents are fortunate enough to have a kid who is so trustworthy, and not all kids are lucky enough to have a parent who cares enough to value relationship over rules. That being said, the sooner we can explicitly communicate our expectations of each other as well as the repercussions if expectations aren't met, the less conflict there will be.

Three years ago, my daughter's public school began a program in which every fourth- and fifth-grade student was issued an iPad as part of the core teaching program. According to the other parents in the school, the first year of the program was fraught with problems because students were chatting by text message in class and staying up till all hours playing on their devices. By the third year, though, the school had learned to set strict boundaries, restricting the types of apps that can be downloaded, removing chat functions, and leading the students through a full week of "digital citizenship" to outline expectations. Students were told in advance that their parents would be monitoring their messages, so that there were no surprises, and since then, the program has been an incredible success.

Identifying these boundary lines might take some trial and error. However, taking the time to set these "invisible fences" or expectations for yourself and for others will save you much trouble in the long run and foster better relationships in your life.

LEARNING NEW BOUNDARIES

Once you decide to build a new fence, recognize that it may take some trial and error and lots of repetition before the boundaries become clear, and even longer before the boundaries are appreciated. Although teens get a bad rap for being tech-obsessed digital natives (young people who have grown up with tech), they are far more savvy and sensitive to tech use than we were at their age. In a focus-group discussion conducted by the Pew Internet & American Life Project in 2013, some particularly self-aware teens indicated that they disliked the increasing adult presence, excessive sharing, and stressful "drama" on Facebook.[50] Sixty percent of teens chose to keep their profiles private, and 74 percent were very comfortable with unfriending or blocking individuals from their feed.

In contrast, those in my generation—and our parents—grew up having free rein with technology because we didn't know any better. I was fourteen years old when America Online (AOL) started shipping out free trial CDs of its dial-up software. On those afternoons when *The Jetsons* wasn't on after school, I had a good hour of time to explore how a modem worked and what the so-called "internet" was. I quickly learned that, while the free trial for the internet was only thirty days, AOL seemed to ship me a new trial CD every month. With just a little bit of hassle, a smidgen of patience, and a whole lot of earsplitting modem noise, I could surf the web to my heart's content, albeit *slowly*. And let me just say, I learned all kinds of things. My parents had no idea, because they were also learning right alongside me (or perhaps several hundred megabytes behind me).

Since that time, this internet has become increasingly sophisticated and nuanced, yet our control of that content has lagged behind. I spoke with John Stix, founder of Fibernetics, one of the largest telecommunications companies in Canada, who has received an increasing number of requests to assist the Royal Canadian Mounted Police in tracking down IP addresses of offenders on the internet.[51] When John heard the story of a six-year-old mentally disabled child who stumbled upon an adult website and suffered from severe guilt and shame, he knew he had to do something. John created a new device called KidsWifi to offer parents a simple, streamlined way of managing content and devices for multiple family members. He explains, "You wouldn't let your child ride a bike without a helmet—so why let them venture on the web unprotected?" Safeguards like KidsWifi help set new invisible fences to get control back into parents' hands—a shift that couldn't happen soon enough.

The new generation of parents is infinitely more savvy about the lure and dangers of technology—we know that boundaries are important for our children, even if we haven't yet set them for ourselves. For instance, we know which apps are good for stretching our children's minds and which apps teach behaviors that we don't like; we can recognize when our children fall off the Happiness Cliff; and therefore, we know when it's time to turn off the devices (whether we actually do that is another question). If we hope to stay a couple megabytes ahead of our kids, we need to actively embrace and test programs that they are using—and as a double benefit, we are able to share a common experience and language with them to help process their changing world.

KNOWING WHEN WE'VE CROSSED THE LINE

Sometimes our device chargers become electronic umbilical cords, which can literally cause separation anxiety when we're disconnected. I remember the first time I saw the black screen of death on my laptop in business school—I felt like I went through the seven stages of grief after the loss of a loved one. If you knew me at that time, you would have thought my entire house had burned down. While I was angry that I'd lost all my notes and projects for school, my dominant emotion was sorrow over the loss of my photos, which were incredibly precious to me. After I emerged from my short-term depression, I became a woman of action who learned to triple back up my most valued documents. And I quit buying the cheapest laptop on the market.

When I look back on that experience, I realize that my initial emotional reaction to my loss was understandable but not acceptable. I don't want an inanimate device to have that much power over me, so I must do something about it. Setting limits on our use of tech is essential not only to our productivity, but also to our health. No one knows this better than Levi Felix, who at the mere age of twenty-four was already creating digital strategy and content for the likes of Jennifer Lopez, Jenny McCarthy, the AARP, Save the Children, and J. K. Rowling. Fueled by passion and a drive for success, Felix landed himself in the hospital with internal bleeding due to what he calls the "perfect recipe of social entrepreneurial-tech burnout: seventy-hour work weeks, stress, and sleeping at the office." After losing 67 percent of his blood and facing near-death, Felix decided to leave his dream job to take a sabbatical. He traded in his digital devices for a backpack, sold everything he owned, and left with a friend to travel the world. Two years, twelve countries, and one ten-day silent retreat later, they returned to start Camp Grounded, a place "[w]here grown-ups go to unplug, get away, and be kids again." They offer camps from one day to one week, complete with fifty-plus playshops for fueling your creativity and inner joy. Sign me up! And actually, I have about fifty teenage Snapchatters I would like to sign up, too. On second thought, maybe I'll sign us up for different camp weeks.

As nice as this sounds, it only offers a short-term fix. I don't think that getting away from tech is a good long-term solution, nor do I think it's something most of us can afford to do! Sometimes stepping away from our devices for even a week is not practical, or even advised. For some professionals like doctors, having access to a phone can be a matter of life or death. Or for a single parent, being unreachable for a long period of time also can be life or death or at least dangerous if there's an emergency. For those who have a desire to step back from technology for a bit, I do think that programs like Camp Grounded can be a breath of fresh air, helping with our impulse-control problems and giving us increased perspective on life. However, as I've been presenting throughout this book, I argue that tech is not a toxin that we need to flush out of our systems—it's a tool. And it's a tool that we must learn to wield effectively.

Tech is not a toxin that we need to flush out of our systems—it's a tool. And it's a tool that we must learn to wield effectively.

Long ago, when the wheel was first invented, I imagine there was a similar explosion of new ideas and possibilities. Wheels were put on everything and they changed the landscape forever. Yes, I am guessing that there were some snafus with objects rolling off the side of cliffs before wheel blocks were invented. Yet, cavemen didn't need a break from the wheel, and no one reverted back to "the good ol' days" before the wheel. Rather, they just learned how to balance the wheel better. As a result, the wheel became a part of life moving forward.

I think if we look at the underlying issue with our increasing tech addiction, there's an elephant in our digital circus, so to speak. And that elephant is that we don't know how to set boundaries in this new uncharted territory of constant connectivity. Perhaps we are running from a boss who won't stop emailing on the weekend, a friend who texts constantly, or an inbox that continually alerts us to how far behind we are on everything. We anthropomorphize our electronics as if they were adversaries pursuing us, but it's worth pausing to think about what is causing that underlying impulse to run from our devices. Life coach Kelci Hart Brock encourages us to chase the fear behind our various behavior triggers. Are you worried that if you don't respond to that text, you might upset someone? Or do you use email as a way to avoid face-to-face confrontation? Do you struggle with checking social media for fear of missing out (FOMO)? Knowing the root of your triggers is the first step to gaining control of your technology.

UNPLUGGING STRATEGICALLY

I walked out of Starbucks looking like a zombie yet again. I had been working on my laptop intensely for a few hours and was sick and tired of staring at my screen and answering an endless stream of emails. I squinted my eyes to adjust to the bright Texas sun and fumbled for my car keys while precariously juggling a Frappuccino in one hand and my laptop in the other. I'm not sure if it was the heat or the caffeine that made me see the mirage, but I envisioned my laptop slipping in slow motion out of my arms and smashing onto the concrete. For just a brief second, I felt this wave of relief, even glee, that I had emerged the victor over some epic struggle between adversaries. But then I immediately thought of the quantities of photos, ideas, and—yes—work documents that I would lose should that happen, and I clutched my computer just a bit tighter.

I drove home in a bit of a daze, trudged into the house, and dropped my laptop bag on the nearest chair. I could tell I wasn't the only one "teched out." One of my bright, beautiful daughters sat with glazed eyes on the couch watching Netflix; another sat next to her mindlessly playing on an iPhone; and the third lounged on a beanbag reading her language arts homework on her iPad with a bored expression (no wonder Steve Jobs never let his children play with these devices!). Meanwhile, my husband was at the kitchen table, busily typing on his work PC, his phone tucked between his cheek and chin, with a serious look on his face.

When no one seemed to acknowledge my presence, I found a plug near the kitchen table and pulled out my laptop to recharge. Where was happiness in the midst of all of this circuitry?

I can tell you, the next several hours in my house did not go well that night. There were tech tantrums (you know, the gnashing of teeth when a device is removed from a miniature zombie), grumpy parents, and late bedtimes. The truth was that the devices were not to be blamed; it was the parenting (ouch, that hurts to admit). In an effort to finagle a little extra work time, my husband and I had inadvertently hired digital babysitters.

I share this vignette from my life because perhaps you can relate. Instead of being plugged in to our devices, sometimes what we need most is to be plugged in to each other. A recent study found that children actually need to power off devices regularly so that they can understand the clear boundaries between the virtual world and the real one.[52] Given that the prefrontal cortex, the region of the brain that controls impulse, does not finish developing until the midtwenties, parents should not be surprised if younger children with smartphones lack impulse control and find themselves more easily addicted.[53] The truth is that all of us need these boundaries in our living and working spaces more than ever. Because maybe there is something abnormal about walking into my home and plugging in my laptop before hugging my husband and kids. We need to learn how to manage our devices, not merely manage to get by.

Rather than just getting away from our devices, I advocate a method I call "strategic unplugging." And to prove to you that tech is a tool and not a toxin, I am going to *use* tech to help you gain greater control over your tech. Our goal is to control our impulses by being intentional about when, where, why, and how we use tech.

Happy Hacks to Get You Started

1. **Know your stats.** Download the Instant or Moment apps to see how many times you turn on your phone each day. The average person checks his phone 150 times every

day. If every distraction took only one minute (a seriously optimistic estimate), that would account for 2.5 hours of distraction every day. That's 912.5 hours a year, or roughly *thirty-eight days each year.* You see the problem? Perhaps even more disturbing is that I recently learned that a leading cellular company now offers its customers five dollars off of their monthly bill if they download an app that enables pop-up ads every time they open their phone, which could exponentially increase distraction. Knowing your stats increases your awareness so that you can make proactive choices about how you spend your time and energy.

> The average person checks his phone 150 times every day. If every distraction took only one minute (a seriously optimistic estimate), that would account for 2.5 hours of distraction every day. That's 912.5 hours a year, or roughly thirty-eight days each year.

2. **Know your limits.** You don't always need to turn off technology—sometimes you just need to learn how to set limits and boundaries so that you do not fall off the Happiness Cliff. As we discussed in Strategy #1, every app we use is subject to the Law of Diminishing Returns, meaning that even the most useful app can be overused. If you can begin to recognize that feeling within yourself when you might be falling off the Happiness Cliff, you can teach yourself to stop the behavior while you are still ahead. To see which apps you might be overusing, download the Break Free app to see how often you use different apps or applications.[54]

   In addition to limiting your phone usage, you can get creative about setting limits for use of technology in other domains of your life, such as abstaining from tech at nighttime, which will improve your productivity and mood for future days, as well. You can also set limits for how many people you follow on Twitter, how many e-books and audio books you buy, or how many apps you own. Rather than trying to consume everything for fear of missing out (FOMO!), learn to introduce only what you can actually consume and enjoy. As my mother always said, "For every new toy you get, you need to give another toy away to make space for it." Bobo observes I conveniently break my iPhone *right* before the new model comes out. *I'm just making space! Wink, wink.* (Let's hope he doesn't get this far in the book.)

3. **Know your weaknesses.** Download the Unplugged app for iPhone or Offtime app for Android to boost your willpower in putting your phone down from time to time. The Unplugged app encourages you to put your phone on airplane mode for short periods of time in order to focus or connect with others better, and the Offtime app whitelists contacts that you want to be able to pierce through your downtime, like your spouse or children, but otherwise shuts down apps, calls, texts, and emails.[55] We know that merely having a phone in your presence—even if you don't touch it—decreases your productivity and weakens your ability to connect with other people.[56,57]

   A recent study of 450 workers in Korea found that individuals who took a short work break without their cellphones felt more vigor and less emotional exhaustion than individuals who toted their cellphones along with them on their breaks, regardless of whether they actually used the phone![58] This could be a great strategy for using your lunchtime to recharge or to connect with friends. In *The Happiness Advantage*, Shawn calls this principle the twenty-second rule because you are trying to increase the activation energy of getting to your phone by twenty seconds, during which time your brain usually decides that that is entirely too much work and gives up (haha, take that, lazy brain!).

4. **Know your intentions.** Download the Live Intentionally app to write explicitly how you would like to use tech in the future. For example, you might write:

   My intention is to use my phone as a tool and not as an escape.
   My intention is to check email only once a day.
   My intention is not to turn on my phone at family dinnertime.
   My intention is to look people in the eye rather than at my screen.

   Without setting an explicit intention for yourself moving forward, the brain will resort to muscle memory and sink into previous habits. Individuals who write down their goals are 42 percent more likely to stick with them.[59] As I mentioned earlier, consider starting your day (weekend days, too) by taking two minutes before you ever touch your phone or computer just to savor an "unplugged space" at the beginning of the day.

Without setting an explicit intention for yourself moving forward, the brain will resort to muscle memory and sink into previous habits. Individuals who write down their goals are 42 percent more likely to stick with them.

A SHARED VISION FOR THE FUTURE

The technological revolution has blurred the lines for human interaction, changing when, why, where, and how we say things. Knowing that technology is here to stay, we have to be active participants in rewriting the rules for how we use technology to communicate and shaping the spaces and places in which we live, work, and learn. And you can start right in your own corner of the world.

> Knowing that technology is here to stay, we have to be active participants in rewriting the rules for how we use technology to communicate and shaping the spaces and places in which we live, work, and learn.

Rather than striving toward a future where the Jetsons represent the gold standard for futuristic living, let's exceed that vision. Let's create a warm future that lifts up the core of humanity by creating space in our lives to live and breathe, by shaping spaces that facilitate the best of human interaction and collaboration, and by building happy fences to foster communication and increase balance in our lives.

A Blog Excerpt from Christine Carter, Author of *The Sweet Spot and Raising Happiness*

Unless you are some sort of superhero, you will not be able to cure yourself of your smartphone/email/internet addiction perfectly the first time. Research indicates that 88 percent of people have failed to keep a new resolution; in my experience as a human being and a coach, 100 percent of people trying to radically reduce their screen time lapse in their attempt. If you struggle with this unplugging business at first, it's important that you don't get too anxious or succumb to self-criticism. Instead, forgive yourself. Remind yourself that lapses are part of the process, and that feeling guilty or bad about your behavior will not increase your future success.

If you aren't succeeding in unplugging, it's important to *figure out what the problem is.* This may be blazingly obvious, but in order to do better tomorrow, you'll need to know what is causing your trip-ups. What temptation can you remove? Were you stressed or tired or hungry when you broke down and checked—and if so, how can you prevent that the next time? Figure it out, and make a specific plan for what to do if you find yourself in a similar situation again. What will you do differently? What have you learned from your slip?

Above all, comfort yourself through this process. To boost follow-through on our good intentions, we need to feel safe and secure. When we are stressed, our brain tries to rescue us by activating our dopamine systems. A dopamine rush makes temptations more tempting. Think of this as your brain pushing you toward a comfort item . . . like the snooze button instead of the morning jog, onion rings instead of mixed greens, or that easy taxi to work rather than the less-than-comfortable urban bike ride. So sometimes the best thing that we can do to help ourselves unplug is to preemptively comfort ourselves in healthy ways before our brain takes matters into its own hands.

SUMMARY

The influx of new technologies into our lives requires both physical and mental space to use these advances properly and in a way that fuels happiness. Strategy #4 focuses on how we can leverage high-impact organizing to regain control of our lives as well as how we can set effective limits and boundaries on our use of technology.

Create a habitat for happiness by:

- ✓ Making space in your life for future happiness by decluttering your environment and your mind
- ✓ Designing with intentionality the places where you live, work, and learn for greater happiness
- ✓ Building happy fences for your use of tech before you fall off the Happiness Cliff

STRATEGY #5 INNOVATE CONSCIOUSLY

HOW TO USE YOUR INNATE POWER TO SHAPE THE FUTURE OF TECHNOLOGY AND HAPPINESS

I remember crying with tears of awe and joy the first time I saw the TED Talk about a ten-year-old boy named Luke Massella, who was born with spina bifida and had already been through sixteen different surgeries in his short life. His bladder was malfunctioning, causing urine to back up in his kidneys and leading his body into kidney failure. Just when things were looking hopeless, Luke's family received a call from Dr. Anthony Atala, who told them about an experimental new surgery to receive a 3D-printed bladder made of synthetic stem cells. The surgery not only worked, but also enabled Luke to flourish and enjoy his life. Today, Luke is a twenty-three-year-old college student who is active on the wrestling team. Dr. Atala explained in his TED Talk that the field of organ development dates to 1938; however, owing to the collaboration of numerous scientists, researchers, innovators, and funders, the Wake Forest Institute for Regenerative Medicine was able to successfully print Luke's bladder, along with many other organs since then.

This is amazing! When I think about the world that my young children will grow up in, the future is incredibly bright and I am hopeful that life will be better for them on so many different fronts. I love seeing stories of #inspirationaltech and continue to share them on social media frequently. These stories of technology being used for good help to supplement the *Jetsons*-like dreams of flying cars that will eliminate traffic and personal robots that can fold our laundry. Yet the million-dollar question is: Will these advances actually make the world a happier or more fulfilled place? If not, are there advances that actually will?

Strategy #5 is a call to action for you to innovate consciously and conscientiously—to step away from how tech can aid in your happiness and instead to move toward how *you* can actively help shape the future of technology to create a happier world. I invite you to join me in thinking about how we want the future to look, how we can leverage existing technology, and how we can use our influence to change the world around us.

THE CHALLENGE:
SEEKING A HAPPINESS UPGRADE

Look around you . . . From your phone to your computer to your car, we have tech in our lives today that would have blown our minds ten years ago. If we could have seen into the future to know that one day Skype would give us the ability to connect instantaneously with friends and family across the globe, surely we would have said that that technology would increase our happiness exponentially. Yet, statistics show that depression rates have doubled over that same ten-year period. Despite incredible advances in every single field of learning, tech has not actually *made* us happier.

Certainly, tech has the *capacity* to influence our happiness levels, but as Shawn writes in *The Happiness Advantage*, "Happiness is a choice—and it's one that we actively have to work toward by cultivating a mindset of optimism and gratitude in the present." Happiness is a means, not an end. It's a tool, not a by-product. It's a mindset, not an outcome. Simply waiting for future tech to make us happy is a setup for disappointment, because every time we upgrade, a new version of happiness comes out.

Happiness is a means, not an end. It's a tool, not a by-product. It's a mindset, not an outcome. Simply waiting for future tech to make us happy is a setup for disappointment, because every time we upgrade, a new version of happiness comes out.

A book by Jacob Weisberg called *We Are Hopelessly Hooked* explains that:

> *Aspirations for humanistic digital design have been overwhelmed so far by the imperatives of the startup economy. As long as software engineers are able to deliver free, addictive products directly to children, parents, who are themselves compulsive users, have little hope of asserting control. We can't defend ourselves against the disciples of captology [tech that captures your attention] by asking nicely for less enticing slot machines.*[1]

Despite feeling a bit depressed just reading the title of Weisberg's book, I refuse to believe that we are "hopelessly hooked," being dragged through the water toward our most certain demise. We can't sit back and hope that the demand-driven economy will nudge us in the right direction—we have to use our innate power to nudge back at the market to tell it where we want it to go through conscious choices.

Twenty years ago, the dominant belief in affective computing (a branch of computer science focused on health and well-being) was that if we want to get smarter, we have to remove emotions that cause biases in decision-making. Rafael Calvo, professor at the University of Sydney and director of the Positive Computing Lab, explains that:

> *If digital technologies are not actively supporting our well-being, it is simply because we have yet to consider it in the design cycle of technology. This oversight has occurred for many reasons, including a historical position*

among engineers and computer scientists that makes us more comfortable staying clear of the difficult-to-quantify and value-laden aspects of psychological impact. In other words, well-being has been not only traditionally overlooked but even consciously excluded from consideration owing to a legacy of industry discomfort.[2]

Yet, we now understand that artificial intelligence is made stronger with emotional intelligence, helping to anticipate risky behaviors and prevent pain. Tech should be a complement, not a replacement, for emotions and thinking, and building a sense of empathy into our tech is just the next frontier. In March 2016, MIT even held a hackathon to design the future of emotion-aware devices, helping to put the "human" back into humanistic design.

This focus on emotions and values used to be perceived as "soft," but is now beginning to proliferate exponentially across disciplines and throughout universities. Joanne Reinhard, an advisor at the UK Behavioral Insights Team, aka "the Nudge Unit," explains that government leaders are now starting to see the benefits of thoughtful-choice architecture design both for delivery of services and for the advancement of a nation. Following the model from the Kingdom of Bhutan, economists, politicians, and policymakers across the globe are now measuring "gross national happiness" in addition to gross domestic product to better capture the health and success of a country.[3] Similarly, corporations such as Apple are beginning to ask of their technology, "Will it make life better? Does it deserve to exist?" A paradigm shift is taking place, not just in technology, but also in the global mindset toward well-being.

Rather than hoping the information superhighway will just magically intersect with Easy Street, let's build a global grid where happiness is in the Master Plan from the start.

Rather than hoping the information superhighway will just magically intersect with Easy Street, let's build a global grid where happiness is in the Master Plan from the start. Let's quit saying, "I wish someone would create an app that does XYZ," or "I wish someone would design a device that can solve this problem"—and stopping there. Instead, let's proactively look for ways that we can make the world a better place and then join our minds together to create solutions.

THE STRATEGY:
CREATING A BALANCED NEW WORLD

If I were to ask you what makes you the happiest in life today, you might say: your family, your friends, your faith. But if I were to ask you what you think the future of happiness looks like, chances are you would spout off something like having a personalized robot to help you manage life (mine would probably have an uncanny resemblance to Bradley Cooper—my husband is fine with this as long as his looks like Kate Upton). However, there's a surprising disconnect in our minds between what we think will make us happy vs. what actually makes us happy, which scientists call an impact bias. And that bias throws us off balance.

There's a surprising disconnect in our minds between what we think will make us happy vs. what actually makes us happy, which scientists call an impact bias. And that bias throws us off balance.

On the other hand, though, we struggle to concoct a vision of the world with good balance between tech and well-being because the challenges of the Digital Era are all new to us. Hollywood has dutifully given us visions of life in a post-apocalyptic world (*The Matrix, Mad Max, Equilibrium, Avatar, WALL-E*), but we lack examples of a positive, non-apocalyptic future. Why? Because this future would be boring—it lacks the central conflict of a dystopian plotline. Yet if the apocalyptic vision is not the type of world we want to live in, then we are going to have to create a different, better future as we go. Rather than creating a Brave New World, as author Aldous Huxley famously wrote about in his forward-thinking 1931 novel, let's set out to create a balanced new world.

Rather than creating a Brave New World, as author Aldous Huxley famously wrote about in his forward-thinking 1931 novel, let's set out to create a balanced new world.

In 1974, French high-wire artist Philippe Petit gained international fame for walking on a tightrope between the Twin Towers of the World Trade Center in New York City. After crossing back and forth over the wire (not just once, but eight times!), he literally lay down on the wire, suspended 1,350 feet in the air over a buzzing crowd of onlookers, and took a little rest. Petit, whose story is now a major motion picture, *The Walk*, explains that during that walk everything around him faded except the wire and himself, and that for the first time in his life he felt truly thankful and at peace.

Petit's experience sounds blissful, if not utterly transcendent. It's the kind of harmonious flow and alignment for which almost every major religious tradition strives and perfectly illustrates my vision for happiness in the Digital Era. Amid the digital circus around us, I envision that each of us might be able to find our sweet spot on the digital wire, balancing tech on one side and happiness on the other. I picture that the strategies in this book might be tools to help us recalibrate when we start to lose our balance. As wire artist Jade Kindar-Martin (who played the stunt double for actor Joseph Gordon-Levitt in *The Walk*) explains, "I know what happens if my foot slips—I know that I have to do this, this, and this. Before I actually fall, there are one hundred technical instances that are going through my mind."

That is the kind of presence of mind that I aspire to have when I use technology. Given that there is no net to the internet, I alone am responsible for my fate. I want to know when I am on the wire (using technology) and when I am starting to lose my sense of balance. I want to know how to compensate or recalibrate in any circumstance so that I not only can survive, but also can actually thrive out there on the wire.

Given that there is no net to the internet, I alone am responsible for my fate.

We have the capacity to become wire artists of the future by using the strategies I've outlined so far in this book: staying grounded (in this case, the closer to the ground, the better), knowing thyself, training our brains, creating a habitat around us for happiness, and innovating consciously. Now, it's time to pivot and think about how *you* can shape the future balance of happiness and tech as a conscious collaborator, conscious consumer, and conscious catalyst.

CONSCIOUS COLLABORATORS

The first way that we can innovate is through conscious collaboration, an idea that I was first introduced to by Kelci Hart Brock, a life coach who activates people to bring their ideas to life. She explains that being a conscious collaborator means more than being a default collaborator or a reluctant collaborator or a collaborator only on paper. Conscious collaboration requires being an active participant in a group that is dedicated to using their ideas, time, and talent to change the world. This is the future of happiness, and I want to share with you four different roles that you can play to make that happen: by contributing to collective knowledge, by creating an impetus for change, by serving as a digital humanitarian, and by shaping the future from your own sphere.

CONTRIBUTING TO COLLECTIVE KNOWLEDGE

In the information economy, collaboration is the ultimate currency. Digital communities are able to perform many tasks more cheaply and quickly than companies or traditional organizations can by "upsourcing" projects to collective wisdom. The strategy of "crowdsourcing," or soliciting ideas and contributions from a large group of people, has transformed the landscape of collaboration.

In his international bestselling book *Wikinomics*, Don Tapscott captures the changing nature of collaboration: "In the last few years, traditional collaboration—in a meeting room, a conference call, even a convention center—has been superseded by collaborations on an astronomical scale."[4] Here are a few well-known examples of how crowdsourcing information through digital collaboration has contributed to collective knowledge for all:

- ✓ **Google Maps**—Since 2005, more than 40,000 individuals have contributed to mapping 28 million miles of road in 194 countries.[5]
- ✓ **Yelp**—More than 69 million users contributed a total of 108 million reviews about local businesses and restaurants in the second quarter of 2016 alone.
- ✓ **Kickstarter**—Since 2011, more than 11 million people have raised $2.6 billion to fund 111,197 projects.[6]
- ✓ **The Climate CoLab**—To date, the Climate CoLab has brought together more than 10,000 people from all over the world to brainstorm solutions to the climate-change problem.[7]

CREATING AN IMPETUS FOR CHANGE

Digital collaborators may act out of a mixture of altruism, ego, or even a desire for fame, but their efforts benefit us all, enabling innovation to get better faster.[8] In Strategy #4, I hinted at the emergence of a new field called Civic Tech that aims to leverage the strength and power of numbers (otherwise known as "social influence") to push for more transparency, efficiency, and accountability in our workplaces, our government, our communities, and our schools. Civic Tech enables the smallest ideas in the most remote parts of the world to gain traction and ignite change. But systems change does not just happen overnight; it requires strong collaborations to help shift deeply entrenched systems.

Recognizing that many of the government's core services (including Veterans Affairs, health care, student loans, and immigration) were using outdated systems, President Barack Obama founded the US Digital Service in 2014 to recruit top tech talent to improve the design and usefulness of the country's most important digital services.[9] Obama went on to explain that "what we realized was that we could potentially build a SWAT team, a world-class technology office inside of the government that was helping agencies. We've dubbed that the US Digital Service . . . and they are making an enormous difference."

I had the opportunity to speak with one of these superstars, twenty-seven-year-old Vivian Graubard, who helped to found the US Digital Service in the White House. As the daughter of a Colombian father and a mother whose parents were Cuban refugees, Vivian had a unique perspective on accessing resources in America. She shared that while some people think of happiness in the future as owning a self-driving Tesla car or a robotic personal assistant, for many people these high-end devices are a luxury, not a necessity. She went on to explain, "Having a Fitbit track your sleeping habits is nice, but knowing that your children and family have a safe place to sleep at night is critical. And sadly not everyone feels that security. For many individuals, having better access to and delivery of basic services like health care or food stamps is the dominant driver of happiness."

While some people think of happiness in the future as owning a self-driving Tesla car or a robotic personal assistant, for many people these high-end devices are a luxury, not a necessity.

Although the federal government spends $84 billion annually on technology, we are not getting $84 billion worth of service—far from it. Many of our systems are outdated, clunky, or downright buggy. The USDS is working through collaborations to overhaul the way that we interact with the government, and Vivian has had a front-row seat for the action.

Though Vivian has assisted with numerous presidential-priority projects, including the Healthcare.gov rescue, her passion is for combatting human trafficking using technology. Vivian said that all too often, budgets, bureaucracy, and lack of communication undercut well-meaning efforts, leaving loopholes for trafficking. If law enforcement has to make a choice between having an officer surf the internet for suspicious posts or putting an officer on the street, the latter would win every time; but there are valuable clues for identifying trafficking rings that we are missing, and this is where technology can help fill in the gaps. In 2012, Vivian helped to launch the Tech vs. Trafficking Project in the White House, which brought together experts from the private sector (like Google and Palantir), nonprofit sector (like FairGirls and Polaris Project), and government agencies (like the Administration for Children and Families). Their mission was to brainstorm low-cost, easy-to-implement solutions to combat human trafficking through collaboration. At the end of the day, they identified three priorities for moving forward: connecting survivors to much-needed services (including treatment options for mental and physical injuries, as well as options for removing scars and tattoos used to brand victims); improving how law enforcement uses technology to comb through large quantities of data; and improving how data is shared across organizations (such as airport security and immigration) to identify victims and eliminate trafficking rings.

For all of the challenges that we face in the Digital Era, collaborations like this one fill me with hope for the future: hope that we will see a breakthrough in crime, hope that increased transparency in the government will lead to better access to services, hope that victims will receive justice and healing. We are just beginning to glimpse what these kinds of changes will mean for government, but we are entering a new frontier in the way that we solve problems. Encouraged by the success of projects like Vivian's, President Obama went on to organize South by South Lawn, an event inspired by South by Southwest, to bring together creators, innovators, and organizers to discuss how we can collaborate better to improve the lives of people around the world.[10]

Cities across the globe are starting to explore how technology can change the infrastructure and delivery of services to more efficiently use public resources. For instance, the City of Barcelona recently installed 1,100 smart lampposts, each of which features Wi-Fi hot spots, LED bulbs that dim when streets are empty, and even sensors that measure air quality. As a result, Barcelona officials now report that energy consumption has decreased by 30 percent, resulting in $37 million in savings to date.[11] The City of Dallas received recognition from the White House in 2015 after launching the Dallas Innovation District, which includes a Living Lab for testing out smart city concepts, an Entrepreneurship Center for seeding new ideas, and the Dallas Innovation Alliance, which brings together public and private partners to scale solutions across the city.[12]

The City of Boston took a slightly different approach to rethinking how technology could transform the urban landscape by launching the "CityScore Dashboard," a public-facing website that grades how the city is performing on everything from fire-department response time to school attendance using a single number (above one is good, below one is bad).[13] By capturing data from different departments in real time, the mayor and his staff can rapidly respond to the city's needs. For instance, when CityScore launched on January 15, 2016, the emergency medical response system was averaging five minutes and fifty-nine seconds from the initial call to getting an ambulance on site. Over the next three months, that response time continued to grow longer, causing the CityScore to drop. Recognizing that there might be a problem, Mayor Marty Walsh reached out to the chief of EMS to get more information on what was going on and learned that the number of visitors and residents in the city had risen over the past several years, which led to an increase in 911 calls; however, EMS's budget had not grown to allow it to hire more EMTs and replace aging ambulances. The mayor was then able to prioritize funding in the next city budget to train twenty new EMTs and buy ten replacement ambulances.[14]

While systemic change on a city, state, or national level might seem overwhelming, it's worth remembering that the world's most powerful collaborations all start with a single person having a single idea and taking a single action. For instance, in the Oak Hills neighborhood of Dallas, Jason Roberts learned that outdated building codes from 1941 were actually fueling blight in the neighborhood, preventing businesses from making improvements to their storefronts (for instance, just placing flowers on the sidewalk required a $1,000 permit). Roberts created a website to recruit volunteers to his cause, aptly named Better Block Initiative.[15] Together, he and his team of volunteers planned a demonstration in Oak Hills for municipal leaders, aiming to beautify the neighborhood, while breaking as many codes as they could in one day. They then invited civic leaders to participate in the events of the day, explaining how many rules they had broken and why. Many leaders were shocked by the existence of these outdated codes, and the simple demonstration led to an overhaul of the city code and the revival of an entire neighborhood. The Better Block initiative has now been replicated in cities across the globe, and continues to gain momentum as best practices are shared through online forums and how-to guides.

The world's most powerful collaborations all start with a single person having a single idea and taking a single action.

SERVING AS A DIGITAL HUMANITARIAN

Another role that you can play as a conscious collaborator is serving as a digital humanitarian. I remember my first exposure to digital humanitarians in June 2005, and the experience was transformative. My husband and I had just moved to Biloxi, Mississippi, to start his first military assignment. At the time, I was still in the middle of business school and I needed to do a business-related internship over the summer. So I decided to intern with the United Way in Gulfport, where I was tasked with helping to update the organization's infrastructure by developing new systems for online giving, emergency protocols, and sustainability planning. I had no idea how soon my efforts would be put to the test. I wish I could say that my work that summer was part of a crucial response initiative, but when Hurricane Katrina hit on August 29, I learned that my protocol, along with my desk and entire filing cabinet, was literally floating in the ocean.

Following my limited memory of my protocol, I called the other staff, only to learn that everyone was in crisis and our executive director was missing (we were able to connect with her a few days later, and she was fine). I remember feeling so helpless—our organization was supposed to be a hub for disaster response in the area, but our operation was completely debilitated. Yet in the interim, digital humanitarian groups from around the globe stepped up, helping to upgrade our infrastructure for communication and filling in where we simply could not.

Since Hurricane Katrina, technology has improved disaster response by leaps and bounds. In times of crisis, digital humanitarians can help sift through the high volume of user-generated content (texts, photos, aerial imagery, videos, and more) so that aid workers can focus on doing what they do best: providing aid. Working in partnership with humanitarian organizations, these volunteers help sift through information, refine priorities, create new structures for operations, and set up websites and communities to provide crucial information. From UNICEF Innovation and UN Global Pulse Jakarta to Ihub Nairobi and Kathmandu Living Labs, numerous organizations offer ways for individuals to lend a hand.

These digital responders use their time and technical skills, as well as their personal networks, to help mitigate information overload for formal humanitarian-aid agencies in the field. For instance, when the Nepal earthquake hit, more than 7,500 volunteers contributed to improve Wikipedia's OpenStreetMap so that aid workers could navigate the area efficiently and even use satellite imagery to determine the regions of Nepal that might be affected. Similarly, Humanity Road and Standby Task Force curate social information and coordinate volunteers, while other groups like Translators without Borders are bridging language gaps both orally and in written form.[16,17]

You don't have to traverse the globe to be a digital humanitarian. Some of the highest-impact work can take place right within your own neighborhood. When I moved into my new neighborhood in Dallas, I learned that a huge percentage of residents use the app NextDoor as a sort of virtual community kiosk. I was amazed to see how this simple, free community board added a depth to our community, deepening connections and enabling a whole different level of social support. On a daily basis, users posted about dogs on the loose, teenagers driving recklessly, families in need, and more. On one memorable day, a man collapsed on a nearby sidewalk and another neighbor found him. He had no identification on him, so the neighbor called 911 and then posted a description of him to alert his family. Within five minutes, his family was found and he was safely on his way to the hospital.

Whether you are serving a community in your own backyard or across the globe, volunteering as a digital humanitarian can be an incredibly powerful form of conscious collaboration.

INNOVATING FROM WITHIN YOUR SPHERE

The beauty of being a conscious innovator is that you don't have to be a computer engineer or a programmer to make a difference—you can innovate right within your own sphere of influence. To demonstrate, I want to tell you the inspiring stories of three individuals who took action in their own communities to create powerful change.

Doc Hendley

Doc Hendley is the unlikely founder of an organization called Wine to Water that aims to provide clean water to those in need around the world. Doc explains, "When the idea came to me to start Wine to Water, the only real job experience I had was tending a bar." The tatted-up, motorcycle-riding bartender had a vision, though. He says, "I dreamed of building an organization that fought water-related death and disease using different methods than anyone else. So, I started raising money to fight this water epidemic the best way I knew how, by pouring wine and playing music."[18] Pretty soon, he was using social media to help individuals around the world turn their cocktail parties into fund-raisers. Since its founding in 2004, Doc's organization has served more than 400,000 people in twenty-four countries. In 2009, Doc Hendley was named one of CNN's Heroes. Today, he speaks with audiences around the globe and says, "If you find something you are passionate about, I don't care who you are, you *will* make a difference."

Michael Pritchard

Michael Pritchard also felt compelled to address the water crisis, but he approached the problem in a completely different way. Michael recalls watching the televised coverage of the twin tragedies of the Asian tsunami and Hurricane Katrina, and feeling saddened by the masses of refugees waiting for a simple drink of clean water. Fueled by a mixture of anger and frustration that aid agencies were so slow to respond, Michael decided to do something about it. He created the Lifesaver bottle, which looks like a sports bottle but includes a nano-filtration membrane (designed to block viruses) to make "even the most revolting swamp water" drinkable in seconds. To prove its effectiveness, Michael gave a TED Talk in which he demonstrated onstage how the bottle worked. He poured runoff from a sewage-plant farm into the Lifesaver, added a smidgen of rabbit poop, shook the bottle, and then drank the clean, sterile water that came out.[19] Michael Pritchard's Lifesaver bottle transformed traditional aid models for providing access to clean water in a faster, more efficient way.

Allie Wilburn

Allie Wilburn, a senior at Nansemond-Suffolk High School in Virginia, became fascinated by the idea of 3D printing during a school course the previous year called Introduction to Design.[20] The following summer, Allie, seventeen, was surfing the internet when she stumbled upon a group called e-NABLE that leverages volunteers to provide 3D-printed prosthetic hands and arms to those who need them. She was further inspired when she got to learn more about prosthetics through some patients at the Hampton VA Medical Center. Allie worked with Elizabeth Joyner, the science, technology, engineering, and math learning and innovation specialist at Nansemond-Suffolk, to land a $50,000 grant to put on a summer robotics academy that culminated in students creating about a dozen prosthetic hands.

All three of these individuals used conscious collaboration to accomplish what no one person can do alone. These civic innovators are without a doubt inspiring, albeit a bit intimidating. We think there is no way that we can have that kind of impact, but as Doc Hendley points out, "My efforts are going to be a drop in the bucket, but if I had never taken that step because it was too big of a problem, then we wouldn't be anywhere right now."

So, what ideas do you have for making the future brighter? Do you aspire to be a civic innovator? If you are interested in collaborating with others on an existing project, here are a few of my favorite apps and websites that you can try out:

Become a Conscious Civic Innovator	
SeeClickFix PublicStuff	Report issues like potholes and street signs that need to get fixed in your city.
Change.org Care2.com	Start petitions and gain support for causes.
NextDoor Peer.by Facebook groups	Build community in your neighborhood.
CitizInvestor.com Neighbor.ly	Invest in public projects through crowdfunding and civic engagement.

CONSCIOUS CONSUMERS

The second way that we can innovate consciously is by being conscious consumers. Buying organic may not be the cheapest option—until, as many argue, you add in the health costs associated with eating too much processed and conventional food. Socially responsible investing may not earn the best return on paper in terms of sheer profit. But we are more than our money. Recent research shows that *how* we spend our money affects our happiness more than *what* we spend our money on.

MINDING OUR DIGITAL FOOTPRINT

In the same way that the environmental movement seeks to raise global awareness about our carbon footprint, so the conscious consumerism movement aspires to raise consumers' awareness about our digital footprint by educating consumers about the impact that their purchases have in the marketplace.[21] Most consumers are completely unaware that the apps we buy, the games we play, the gadgets in which we invest, and the conversations we have on social media all send messages to developers and investors about the kind of content we value most and the products that we desire. Yes, data drives dollars—and consumers have an incredible power to influence market forces.

FOSTERING SOCIAL RESPONSIBILITY

Tech has become a catalyzing force for positive change, increasing consumer knowledge and corporate transparency. Today's consumers are more aware of corporate practices and know that they have options about where to spend their money. According to the 2015 Cone Communications survey, eight out of ten global consumers consider corporate social responsibility when deciding what to buy or where to shop (84 percent), which products and services to recommend to others (82 percent), which companies they want to see doing business in their communities (84 percent), and where to work (79 percent).[22] As a direct result, 93 percent of the world's largest companies now formally report on corporate social responsibility (CSR).[23,24]

Eight out of ten global consumers consider corporate social responsibility when deciding what to buy or where to shop (84 percent), which products and services to recommend to others (82 percent), which companies they want to see doing business in their communities (84 percent), and where to work (79 percent).

Being a conscious consumer is not only good for the world, but it can also be good for you. Feeling good about a purchase not only makes you happy in the short run, it can also cause you to double down on altruism in the future.[25] Research suggests that there is a "positive feedback loop" between prosocial spending (spending for the good of others) and well-being that could be a powerful path toward sustainable happiness: prosocial spending increases happiness, which in turn encourages further prosocial spending. If you want to be a more socially responsible consumer, here are some very simple ideas you can try.

Happy Hacks to Get You Started

1. **Source your products.** Know where your products come from, who makes them, and how profits are used. Apps like GoodGuide, Ecohabitude, and Buycott can help you rapidly scan a barcode to check into a company's background before purchasing a product from it.[26] If a barcode is not available, you can also look the product or company up on Google or the Better Business Bureau.[27]
2. **Buy for impact.** When possible, buy products from companies that support a triple bottom line (meaning that the purchase benefits the economy, environment, and community).[28] Usually these companies embrace a corporate social responsibility plan, and they may even be certified as a "B Corp," indicating that they meet the highest standards of verified social and environmental performance, public transparency, and legal accountability.[29]

   While these companies might sacrifice short-term profits for their values, they are building brand loyalty among consumers who want to know where their dollar is going and increasing their recruitment potential for top talent. Kristen Drapesa, founder and CEO of EcoHabitude, a socially responsible online marketplace, entreats, "The sooner we start to vote better with our dollars, the faster we will see the shift in the economy . . . It's all about demand, and we're starting to see that shift in what people are looking for and care about."[30]
3. **Invest in innovation for the common good.** The next best thing to inventing a solution to make the world a better place is investing in an inventor. Whether you have one dollar or a million dollars, today you can be a philanthropist thanks to crowdfunding sites like Kiva, Kickstarter, Crowdrise, and Indiegogo. Investing in projects that align

with your values and interests helps to generate actionable capital for positive change, signaling to other inventors and investors that there is demand for similar projects and products.

4. **Use social media as a platform for change.** Celebrate companies that are committed to making a positive difference, and push back against corporations that demonstrate breaches of integrity. Post product reviews, engage in message boards, and write blogs to express yourself. You can also follow social media pages for organizations that align with your values to stay more informed about key issues.

CONSCIOUS CATALYSTS

The third way that we can innovate consciously is by playing the role of catalyst in society by modeling, transforming, nudging, and giving. In most cases, our actions serve a triple purpose, improving our own happiness, inspiring others to action, and setting up an environment that fosters and facilitates greater happiness for all.

MODELING PROSOCIAL BEHAVIORS

By modeling prosocial behaviors, innovators have the ability to be catalysts in society for increased altruism, empathy, and reciprocity. In the famous 1966 Flat Tire experiment, researchers studied whether or not motorists would stop to lend a hand to a "lady in distress" stopped on the side of the road with a flat tire. Half of the drivers had seen a staged setting with a young male helping a girl just beforehand, while the other half of the drivers had not. The study found that the presence of a positive model significantly increased the altruistic behaviors of other drivers.[31] In fact, observing an act of kindness can set a cascade of generosity into motion. In 2012, a customer at the drive-through window of a Tim Horton's coffee shop decided to pick up the tab for the stranger in the car behind her in line. Surprised and delighted, the customer behind her decided to pay for the following customer as well, resulting in a 226-customer streak of generosity over the next three hours.[32] A similar trend occurred in 2014 at a Starbucks drive-through window, resulting in a 378-customer streak over eleven hours. In each of these cases, a single act of altruism created a powerful ripple effect that extended far beyond the people in line—these stories became an internet sensation and an inspiration to others to be catalysts in their communities. (To track your impact, check out the Ripil or Nobly apps, which enable you to track and share the good you create in the community around you.)

The study found that the presence of a positive model significantly increased the altruistic behaviors of other drivers. In fact, observing an act of kindness can set a cascade of generosity into motion.

Whether you are a CEO, a team leader, a front-line worker, or a stay-at-home parent, behavior modeling can be an extremely effective way of motivating individuals around you. In his *New York Times* best-seller *Give and Take*, Wharton professor Adam Grant writes, "The greatest untapped source of motivation is a sense of service to others; focusing on the contribution of our work to other people's lives has the potential to make us more productive than thinking about helping ourselves."

Grant walks the talk: he once agreed to write 100 recommendation letters for students, a task that many professors might find onerous. Rather than thinking of the letters as a chore, though, he opted to think of each one as an opportunity to help others reach their dreams. While this sounds Pollyanna-ish, Grant acknowledges that this mindset shift was not purely altruistic—he got a lot out of the deal, as students regularly brought him gifts, sent him thank-you emails, and spread the word about his character.

If you still doubt the value of giving over taking in the office, Grant goes on to explain a simple experiment he tried to demonstrate the power of framing on work performance, starting with a notoriously difficult environment: campus call centers. Since the call center's primary purpose was to fund scholarships, Grant brought in a previous scholarship recipient to talk to the callers for just ten minutes about how much the scholarship had changed his life, inspiring him to be a teacher with Teach for America. Grant recalls that even he was shocked by how effective the experiment was: a month after the testimonial, workers were averaging 142 percent more time on the phone and revenue had gone up 171 percent.[33]

Modeling prosocial behaviors through testimonies and actions does more than inspire, though; it actually changes the hormones in our bloodstream and shapes our behavior on a neurophysiological level. In one study conducted in December 2015, researchers asked subjects to watch a public service announcement; before watching the video, half of the subjects were given oxytocin (a hormone that is released into the blood when one is petting one's dog, attending a wedding, having sex, or watching an emotional YouTube clip), while the other half were given nothing. Participants who received oxytocin donated 56 percent more money, gave to 57 percent more causes, and reported 17 percent greater concern for those in the ads. While we have known innately that "tugging on our heartstrings" is an effective strategy for advertising, this research reveals that modeling "feel-good" activities actually changes the behaviors of onlookers by nudging them toward positive choices.[34]

Moreover, altruistic behavior improves our work outcomes. A GoodThink research study found that individuals who provide social support to others are ten times more engaged and have a 40 percent higher likelihood of promotion over the next four years. Understanding how to leverage this neurophysiological response can be one of the most untapped sources of charisma in leadership, and those who understand how to wield this influence will undoubtedly be more revered and effective.

TRANSFORMING THE NEWS

Innovators also can be catalysts through social media. In a controversial experiment, Facebook altered the news feeds of more than a million users to reflect either more positive posts or more negative posts.

Although the method was suspect, the results of the study were powerful: seeing positive posts in a newsfeed influenced people to post positive updates themselves, whereas seeing negative posts influenced people to post more negative updates.[35]

We literally become what we read. Knowing this, journalist Rashanah Baldwin aimed to change the news coverage of her hometown, Englewood, IL, a neighborhood of Chicago. Englewood is often portrayed by the media as a hotbed of crime, yet Baldwin knows there is more to the story and is committed to sharing news about the good side of her hometown with the world. In one of her articles she explains that, "The danger behind a single, negative theme being told over and over again to the masses, about a single community, will no doubt negatively influence those who live in that community—as well as everyone else. Bottom line: You start to become the perception, if there's nothing to counter it."[36]

Baldwin's approach is what Michelle Gielan, a former national CBS News anchor turned positive psychologist (also my sister-in-law), calls "transformative journalism." In her bestselling book, *Broadcasting Happiness,* Gielan defines transformative journalism as "an activating, engaging, solutions-focused approach to covering news. It seeks to inform the public while providing the necessary tools to create forward progress." Gielan notes that media coverage is too often siloed into "positive" and "negative" stories, but the real story is more complex than that. "Transformative journalism is the third path that says 'we're going to cover serious news but we're going to cover it in a way that leaves you feeling like your actions matter.' It provides potential calls to action so that people can feel like they're part of the process. It has a greater goal than just telling you bad stuff in the world."

Transformative journalism is not only more palatable, it's also more engaging. In a news media study that Gielan conducted with Arianna Huffington and Shawn Achor (who also happens to be her husband), Gielan found that when people are exposed to a discussion of potential or actual solutions following news of a problem, creative problem-solving on subsequent unrelated tasks increases on average by 20 percent. People also report feeling better after they focus on the solutions instead of merely getting stuck on the problems. This reminds us that fostering a sense of empowerment about one issue can positively influence how we feel about and approach other areas of life.

In a previous study, the same trio found that people exposed to just three minutes of negative news first thing in the morning have a 27 percent higher likelihood of reporting their day as unhappy six to eight hours later. Switching up negative stories with transformative ones makes people feel better, because they see that they have the power to make this world a better place. This is a profound impact, especially for those of us who like to wake up by reading the top five news stories before even getting out of bed. Gielan explains that the goal is not for you to stop watching the news, but rather for you to opt in to transformative news sources that guide you in how to process negative events and act upon them. If these sources are not available, you can always actively read news stories through a lens of what you can do about the situation and "walk the talk" by engaging on social media to shape the conversation with solutions-focused posts and blogs.

NUDGING POSITIVE BEHAVIORS

The third way that innovators can be conscious catalysts is by nudging positive behaviors through technology. Tristan Harris, a graduate of Stanford's Persuasive Tech program and former Google engineer, is fascinated with how technology can nudge human values, such as the notion of "time well spent," in the design of consumer technology. For example, he envisions a version of Gmail that might open by asking you how much time you want to spend on email that day and remind you when you're nearing the limit. In addition, he wants to reengineer messaging apps to value attention over interruption. He calls these "thoughtful apps," and they are essentially apps to control our apps.

He envisions a version of Gmail that might open by asking you how much time you want to spend on email that day and remind you when you're nearing the limit.

As we raise our consciousness about the interplay of technology in our lives, each of us has a responsibility to shape the future of happiness in the ways in which we collaborate, consume, and act as catalysts. As President Obama wrote in a recent edition of *WIRED* magazine, in order to overcome the challenges we face in the future ahead:

> *We need not only the folks at MIT or Stanford or the NIH but also the mom in West Virginia tinkering with a 3D printer, the girl on the South Side of Chicago learning to code, the dreamer in San Antonio seeking investors for his new app, the dad in North Dakota learning new skills so he can help lead the green revolution. That's how we will overcome the challenges we face: by unleashing the power of all of us for all of us. Not just for those of us who are fortunate, but for everybody.*[37]

This is the power of the creative, collective imagination. My goal throughout this book has been to empower you to reflect on where you have been, to think about where you are going, and to dream about what the future might look like. I'm optimistic that no matter what challenges we face, we can, and we will, find a better way forward. I look forward to the future of happiness and I will see you there.

SUMMARY

The future of happiness is directly shaped by how we interact with technology today. Our ideas, actions, and even our purchases send important messages to investors and developers of new technologies. We can help co-create this future by being conscious and intentional about how we engage with technology in the present.

Innovate consciously by:

- ✓ Actively envisioning the world you want
- ✓ Leveraging existing technology to be a conscious collaborator, a conscious consumer, and a conscious catalyst
- ✓ Using your influence to model prosocial behaviors, create a ripple effect, and broadcast positive stories

A CONCLUDING LETTER

My mic drop for the end of the book was going to be a fancy 3D holographic picture of yours truly with two thumbs up congratulating you on finishing this book. But sometimes to move forward, we need to take a look backward. So instead, I'm going to go a little old school. Really old school: I've written you a letter to send you on your way. Twitter and text fiends, this is probably going to be painful for you (way more than 140 characters, no emoticons . . . u will be OK, lol), but I'll try to keep it brief.

Dear Friend (and you are indeed my friend after making it through all of my stories, lists, and Happy Hacks),

I don't believe in long conclusions, especially because this is just a beginning. But I want to say thank you. Thank you for caring enough about the future to join me on this journey of evoking potential, digging into research, and engaging in a thought experiment to make the future just a little bit brighter. Over the course of this book, we've explored the challenges of the modern juggling act, and I have shared with you five strategies to help you balance productivity and well-being in the Digital Era, by:

Staying grounded
Knowing thyself
Training your brain
Creating a habitat for happiness
Innovating consciously

By now, you've probably already identified which of these strategies that you most want to work on, and my hope is that you will start to put these practical suggestions into action right now. Not tomorrow, or next week, or after you've had a chance to buy a gratitude journal. These strategies require nothing but an open mind and an earnest desire to shape the world around you for the better.

If you're reading this book, and the future is still the future, then good. There's still time to strive toward balancing tech and happiness. If you're reading this and the future is now the present, then I need to write another book! Or I need to travel back to the past and make some serious edits.

If you would like to continue learning about cutting-edge resources and strategies for balancing productivity and well-being in the Digital Era, please check out my HappyTechBlog.com. If you find yourself failing, flailing, or falling along the way, take comfort: we are in this together. Read your favorite parts of the book as a refresher. And then, share the story of your success with me at amyblankson.com/share. I can't wait to hear from you and to see how together we can continue to shape the future of happiness.

Yours truly,

Amy

P.S. And, when I'm not on a digital vacation, you can find me @amyblankson on Facebook, Twitter, and Instagram.

SUMMARY OF KEY POINTS

STRATEGY #1: STAY GROUNDED

How to Focus and Channel Your Energy with Intention

Use a QR-reader app to download this summary.

Although our attention spans might be shorter than those of goldfish, we can learn to become less distracted and more present in our lives, so that we can tap into a greater sense flow, fully immersed and engaged in whatever activity we might be doing. We can increase our awareness and intentionality to mindfully choose when, where, why, and how we engage with technology. These skills make up the third prong with which we can ground ourselves and channel our energy toward creating a happier future.

Stay grounded in the midst of change by:

- ✓ Utilizing the "third prong" (your guiding principles and values) to focus your energies
- ✓ Hacking distractions to increase productivity
- ✓ Actively choosing how you want to respond to technology: resist, accept, or embrace it
- ✓ Understanding others' intentions as well as your own
- ✓ Focusing on tuning in, not zoning out
- ✓ Bringing your priorities to the foreground

STRATEGY #2: KNOW THYSELF

How Quantifying Yourself Helps Eliminate Limiting Beliefs on Your Potential

Self-knowledge is power—paying attention and giving intention to the microdecisions in our lives helps us to avoid limiting beliefs and reach more of our potential. New technology helps us to understand our own bodies and minds on an incredibly detailed level, enabling us to make better microdecisions about the future. However, we still have to tap in to the greatest supercomputer ever created, the human mind.

Know thyself by:

- ✓ Learning to recognize limiting beliefs that might derail your best intentions
- ✓ Magnifying your microdecisions to achieve more of your potential
- ✓ Tracking progress in your life to determine where you have succeeded and where you have room for improvement

STRATEGY #3: TRAIN YOUR BRAIN

How to Put Together the Building Blocks of a Smarter, Happier Mind

Recent advances in positive psychology reveal that we can train our brains to improve happiness and performance by using the S.T.A.G.E. framework. To create sustainable positive change in our lives, the key is to identify a target skill set, hone in on one habit, assess your progress, and make the change "sticky" by setting simple, relevant, and realistic goals.

Train your brain by:

- ✓ Developing an optimistic mindset to fuel your growth
- ✓ Using the S.T.A.G.E. framework (Savor, Thank, Aspire, Give, and Empathize) to learn skills for improving your mindset
- ✓ Tapping in to technology to bolster your success and track your progress

STRATEGY #4: CREATE A HABITAT FOR HAPPINESS

How to Build Greater Happiness into Our Homes, Workplaces, and Communities

The influx of new technologies into our lives requires both physical and mental space to use these advances properly and in a way that fuels happiness. This strategy focuses on how we can leverage high-impact organizing to regain control of our lives, as well as how we can set effective limits and boundaries on our use of technology.

Create a habitat for happiness by:

- ✓ Designing the places that we live, work, and learn in for greater happiness

- ✓ Making space in your life for future happiness by decluttering your environment and your mind
- ✓ Setting limits on your use of tech before you hit the Point of Diminishing Returns

STRATEGY #5: BE A CONSCIOUS INNOVATOR

How to Use Your Innate Power to Shape the Future of Technology and Happiness

The future of happiness is directly shaped by how we interact with technology today. Our ideas, actions, and even our purchases send important messages to investors and developers of new technologies. We can help co-create this future by being conscious and intentional about how we engage with technology in the present.

Innovate consciously by:

- ✓ Actively envisioning the world you want
- ✓ Leveraging existing technology to be a conscious collaborator, a conscious consumer, and a conscious catalyst
- ✓ Using your influence to model prosocial behaviors, create a ripple effect, and broadcast positive stories

GLOSSARY

Affective computing—the study and development of systems that can recognize, interpret, process, and simulate human emotions

Civic technology—digital tools used to support, create, or facilitate the public good[1]

Collisionable spaces—places that can lead to serendipitous encounters through intentional office design and work flows

Continuous partial attention—the process of paying simultaneous attention to a number of sources of incoming information, but at a superficial level[2]

Digital humanitarians—individuals who provide online response and relief services to communities affected by disaster events[3]

Dopamine—a neurotransmitter that helps control the brain's reward and pleasure centers[4]

Fixed mindset—a state of mind in which people believe that their basic qualities, like their intelligence or talent, are fixed traits at birth[5]

FOMO—the "fear of missing out" on events[6]

Gamification—the concept of applying game mechanics and game-design[7] techniques to engage and motivate people to achieve their goals

Gestalt psychology—a branch of traditional psychology that tries to understand how we acquire and maintain meaningful perceptions in an apparently chaotic world

GOT (Guilt Over Things) Syndrome—the feeling that we have *got* to keep things, even if we haven't touched them in weeks, months, years—or even decades

Growth mindset—a state of mind in which people believe that their most basic abilities can be developed through dedication and hard work[8]

Happiness—the joy we feel striving after our potential

Happiness advantage—an optimal state where the brain performs better at positive than it does at negative, neutral, or stressed[9]

Happiness Cliff—a ledge that represents the furthest limit at which we experience joy from a diversion; often we become so engrossed in a task that there is a time lag before we realize that our diversions are no longer making us happy anymore

Illusory knowledge—information that helps us to make quick sense of the world, even though there may be gaps that lead to faulty assumptions

Law of Prägnanz—a principle of Gestalt theory that says people will perceive and interpret ambiguous or complex images as the simplest forms possible[10]

Lifehacks—shortcuts to increase productivity and well-being in one's life

Lifelogging—a method of studying the world around you over time by tracking personal data

Limiting beliefs—thoughts that hold you back from achieving your full potential

Microdecision—a small choice you make daily that has a big impact on your life

Mindprint—a unique amalgamation of one's thought processes, intentions, goals, and interests

Moore's Law—a prediction made in 1965 by Gordon Moore, co-founder of Intel, that the number and speed of microchips would double every two years or so

More Than Moore's Law—a prediction that the computing industry will use applications to determine what chips are needed to support them in the future[11]

Optimism—the belief that our behavior matters

Oxytocin—a powerful hormone that acts as a neurotransmitter in the brain; also known as "the cuddle hormone"

Persuasive technology—a vibrant interdisciplinary research field, focusing on the design, development, and evaluation of interactive technologies aimed at changing users' attitudes or behaviors through persuasion and social influence, but not through coercion or deception[12]

Law of Diminishing Returns—the point at which the benefit gained is less than the amount of time or energy invested[13]

Prosocial behaviors—voluntary behaviors intended to benefit another[14]

Quantified self—a movement to incorporate technology into data acquisition on aspects of a person's daily life in terms of inputs[15]

"Really?!" Rule—a litmus test for whether to keep an item, in which one asks oneself, "Does this tech truly make me happier and/or more productive?"

Stereopsis—the perception of depth produced by the reception in the brain of visual stimuli from both eyes in combination[16]

Transformative journalism—an activating, engaging, solutions-focused approach to covering news[17]

TECHNOLOGY REFERENCED

1password	1password.com
Addapp	Addapp.io
AffdexMe	Affectiva.com
Anonymous	Anonofficial.com
BeHppy	Behppy.com
Better Block	Betterblock.org
BreakFree	Breakfree-app.com
Buycott	Buycott.com
Campfire	Campfireapp.io
Care2	Care2.com
Change	Change.org
Charity Miles	Charitymiles.org

Chronos	Getchronos.com
CitizInvestor	Citizinvestor.com
CityScore	Boston.gov/cityscore
Crowdrise	Crowdrise.com
Deedtags	Deedtags.com
Donate a Photo	Donateaphoto.com
ECHO	Kenzen.com
Ecohabitude	Ecohabitude.com
e-NABLE	Enablingthefuture.org
e-Stewards	E-stewards.org
Feedie	Wethefeedies.com
Fitbit	Fitbit.com
GiveGab	Givegab.com
GiveMob	Givemobapp.org
Global Digital Citizen Foundation	Globaldigitalcitizen.org
Goodguide	Goodguide.com
Gratitude 365	Gratitude365app.com
Gratitude Journal	Getgratitude.co
H1	Halowearables.com
HabitBull	Habitbull.com

Habitica	Habitica.com
Habitify	Habitify.co
Happier	Happier.com
Happify	Happify.com
Headspace	Headspace.com
Home	Apple.com/ios/home
Humanity Road	Humanityroad.org
Humans of New York	Humansofnewyork.com
Icis	Laforgeoptical.com
iMotions	iMotions.com
Indiegogo	Indiegogo.com
Insight Timer	Insighttimer.com
Instant	Instantapp.today
Instead	Instead.com
ipassword	iPassword.com
iWatch	Apple.com
Jawbone	Jawbone.com
Kickstarter	Kickstarter.com
KidsWifi	Kidswifi.com
Kiva	Kiva.org
Koto Air	Koto.io

Laster SeeThru	Laster.fr/products/seethru
Life360 (formerly Chronos)	Life360.com
LifeSaver	Iconlifesaver.com
LifeSum	Lifesum.com
LiveHappy	Livehappy.com
Live Intentionally	Liveintentionallyapp.com
Lumo Lift	Lumobodytech.com
MapMyRun	Mapmyrun.com
Meta	Getameta.com
Mindfulness Training	Mindapps.se
Moment	Inthemoment.io
Moodies	Beyondverbal.com
MoodMeter	Moodmeterapp.com
Muse Headband	Choosemuse.com
MyFitnessPal	Myfitnesspal.com
Narrative Clip	Narrativeapp.com
Neighbor.ly	Neighbor.ly
Nest	Nest.com
NextDoor	Nextdoor.com
Nobly	Nobly.com
Norm-Social Philanthropy	Thenormapp.com

Oculus	Oculus.com
Offtime	Offtime.co
One Today	Onetoday.google.com
OpenStreetMap	Openstreetmap.org
OpenTime	Opentimeapp.com
Peerby	Peerby.com
Plasticity Labs	Plasticitylabs.com
Potentia	Potentialabs.com
Productive	Productiveapp.io
Project Aura	Projectaura.com
PublicStuff	Publicstuff.com
Quantified Self	Quantifiedself.com
RealizD	Realizd.com
Remindfulness	Remindfulnessapp.com
Ripil	Ripil.com
Roomba	iRobot.com
Runcible	Mono.hm
SeeClickFix	Seeclickfix.com
Shutterfly	Shutterfly.com
Snapchat Spectacles	Spectacles.com
Soul Pancake	Soulpancake.com

Spare	Sparenyc.org
Spire Stone	Spire.io
Standby Task Force	Standbytaskforce.org
Translators without Borders	Translatorswithoutborders.org
Unplugged	Unpluggedtime.com
Upright	Uprightpose.com
Way of Life	Wayoflifeapp.com
Wine to Water	Winetowater.org
Yale Online	Coursera.org/yale

NOTES

Introduction

1 Surowiecki, James. "Technology and Happiness." *MIT Technology Review*. 2005. <https://www.technologyreview.com/s/403558/technology-and-happiness/>

2 Report. "US Patent Statistics Report." US Patent and Trademark Office, 2015. <https://www.uspto.gov/web/offices/ac/ido/oeip/taf/us_stat.htm>

3 Report. "Digital Eye Strain." *The Vision Council*. 6 January 2016. <https://www.thevisioncouncil.org/digital-eye-strain-report-2016>

Where Are We Heading?

1 "Information about Sea Turtles: Threats from Artificial Lighting." *Sea Turtle Conservancy Home Page*. Accessed 3 February 2016. <http://www.conserveturtles.org/seaturtleinformation.php?page=lighting>

2 "What Is STOP?" *Sea Turtle Oversight Protection About Page*. Accessed 3 February 2016. <http://www.seaturtleop.com/index.php/what-is-s-t-o-p>

3 Report. "The Sea Turtle Friendly Lighting Initiative." *University of Florida Law Clinics Page*. April 2014. <https://www.law.ufl.edu/_pdf/academics/clinics/conservation-clinic/Legal_and_Biological_Introduction.pdf>

4 Weller, Chris. "Exposure to Artificial Light from Electronics Disrupts Sleep Pattern, Causes Decreased Melatonin and Difficulty Falling Asleep." *Medical Daily Home Page*. 1 July 2013. <http://www.medicaldaily.com/exposure-artificial-light-electronics-disrupts-sleep-pattern-causes-decreased-melatonin-and-247286>

5 Gamble, Amanda L., Angela L. D'Rozario, Delwyn J. Bartlett, Shaun Williams, Yu Sun Bin, Ronald R. Grunstein, and Nathaniel S. Marshall. "Adolescent Sleep Patterns and Night-Time Technology Use: Results of the Australian Broadcasting Corporation's Big Sleep Survey." *PLoS ONE* 9, no. 11 (2014). doi:10.1371/journal.pone.0111700.

6 "Could You have 'Text Neck' Syndrome?" *HealthXchange Home Page.* Accessed 5 March 2016. <http://www.healthxchange.com.sg/healthyliving/HealthatWork/Pages/Could-You-Have-Text-Neck-Syndrome.aspx>

7 "Psychophysiological Patterns during Cell Phone Text Messaging: A Preliminary Study." *Applied Psychophysiology and Biofeedback*. 1 March 2009. <http://link.springer.com/article/10.1007/s10484-009-9078-1>

8 Sloane, Matt. "Text Neck' and Other Tech Troubles." Pain Management Health Center. *WebMD Home Page.* 26 November 2014. <http://www.webmd.com/pain-management/news/20141124/text-neck>

9 Cuddy, Amy. "Your iPhone Is Ruining Your Posture—and Your Mood." *New York Times*. 12 December 2015. <http://www.nytimes.com/2015/12/13/opinion/sunday/your-iphone-is-ruining-your-posture-and-your-mood.html?_r=0>

10 Renes, Kevin. "Abstract Illustration of Evolution." *Shutterstock Home Page.* Accessed 16 September 2016. <http://www.shutterstock.com/pic-70386574.html>

Would We Be Better without Tech?

1 Stieben, Danny. "Google Glass Review and Giveaway." *Makeuseof.com*. 23 December 2013. <http://www.makeuseof.com/tag/google-glass-review-and-giveaway/>

2 *Crucial.com Home Page.* Accessed 5 March 2016. <http://www.Crucial.com>

3 Goldman, David. "The Hottest Tech of 2015 and Beyond." *CNN Money–US*. Updated 21 March 2012. <http://money.cnn.com/galleries/2012/technology/1203/gallery.coolest-tech-2015/7.html?iid=EL>

4 Tucker, Abigail. "How to Become Engineers of Our Own Evolution." *Smithsonian.com*. 1 April 2012. <www.smithsonianmag.com/science-nature/how-to-become-the-engineers-of-our-own-evolution-122588963/?no-ist>

5 Bruce, James. "Four Technologies That Could Change the World." *MakeUseOf.com*. 30 October 2011. <http://www.makeuseof.com/tag/4-technologies-change-world/>

6 Al-Rodhan, Navyef. "Inevitable Transhumanism? How Emerging Strategic Technologies Will Affect The Future of Humanity."

The CSS Blog. 29 October 2013. <http://isnblog.ethz.ch/security/inevitable-transhumanism-how-emerging-strategic-technologies-will-affect-the-future-of-humanity>

7 Kushlev, Konstadin. "Checking Email Less Frequently Reduces Stress." Computers in Human Behavior. *ScienceDirect.com.* 1 February 2015. <http://www.sciencedirect.com/science/article/pii/S0747563214005810>

8 Calvo, Rafael A., and Dorian Peters. *Positive Computing: Technology for Wellbeing and Human Potential.* Cambridge, MA: MIT Press, 2014.

9 Gribetz, Meron. "A Glimpse of the Future through an Augmented Reality Headset." *Ted.com.* 1 March 2016. <https://www.ted.com/talks/meron_gribetz_a_glimpse_of_the_future_through_an_augmented_reality_headset/transcript?language=en#t-1072>

What Will Happiness Look Like?

1 Aristotle. *Nicomachean Ethics*. Rev. ed. Edited by H. Rackham. Loeb Classical Library. Cambridge, MA: Harvard University Press, 1934.

Strategy #1: Stay Grounded

1 "Technology Addiction, Concern, and Finding Balance." *CommonSense.com.* Accessed 15 March 2016. <https://www.commonsensemedia.org/research/technology-addiction-concern-controversy-and-finding-balance>

2 Art Markman. "How Distraction Can Disrupt You." Ulterior Motives. *PsychologyToday.com.* 18 March 2014. <https://www.psychologytoday.com/blog/ulterior-motives/201403/how-distraction-can-disrupt-you>

3 Madden, Mary, and Amanda Lenhart. "Teens and Distracted Driving: Texting, Talking and Other Uses of the Cell Phone Behind the Wheel." *Pew Research Center.* 16 November 2009. <http://www.pewinternet.org/2009/11/16/teens-and-distracted-driving/>

4 Sullivan, Bob, and Hugh Thompson. "Brain Interrupted." *New York Times Sunday Review.* 4 May 2013. <http://www.nytimes.com/2013/05/05/opinion/sunday/a-focus-on-distraction.html>

5 *LasterTechnologies Home Page.* Accessed 13 September 2016. <http://laster.fr/products/seethru/>

6 *LaForgeOptical Home Page.* Accessed 13 September 2016. <http://www.laforgeoptical.com/>

7 "Avoid the Unexpected." *Google Glass Home Page.* 28 May 2016. <https://developers.google.com/glass/design/principles#avoid_the_unexpected>

8. Smith, Aaron. "The Best (and Worst) of Mobile Connectivity." *Pew Research Center.* 30 November 2012. <http://www.pewinternet.org/Reports/2012/Best-Worst-Mobile.aspx>
9. Kelleher, David. "Survey: 81% of US Employees Check Their Work Email Outside of Work Hours." *TalkTechToMe Home Page* 21 May 2013. <http://www.gfi.com/blog/survey-81-of-u-s-employees-check-their-work-mail-outside-work-hours/>
10. Meeker, Mary, and Liang Wu. "Internet Trends D11 Conference." 29 May 2013. <http://www.slideshare.net/kleinerperkins/kpcb-internet-trends-2013>
11. "It's Happening Now: Preorder a Runcible of Your Very Own Today." *Monohm Home Page.* Accessed 15 September 2015. Runcible. <http://mono.hm/runcible.html>
12. "Runcible—Circular Open Source Anti-Smartphone." *Indiegogo Home Page.* 22 July 2016. <https://www.indiegogo.com/projects/runcible--2#/>
13. Kelly, Heather. "A $499 Phone for People Who Hate Phones." *CNN Money Home Page.* 15 June 2016. <http://money.cnn.com/2016/06/15/technology/runcible-phone/index.html>
14. Solow, Robert. "We'd Better Watch Out." *New York Times Book Review.* 12 July 1987, p. 36.
15. Iny, Alan, and Luc de Brabandere."The Future Is Scary. Thinking Creatively Can Help." *BCG Perspectives.* 9 October 2013. <https://www.bcgperspectives.com/content/commentary/innovation_future_scary_thinking_creatively_help/>
16. Gerdeman, Dina. "How Electronic Patient Records Can Slow Doctor Productivity." 26 March 2014. *Harvard Business School: Working Knowledge.* <http://hbswk.hbs.edu/item/how-electronic-patient-records-can-slow-doctor-productivity>
17. Granderson, L. Z. "Do You Remember Your Mom's Phone Number?" *CNN Opinion Home Page.* 30 September 2012. <http://www.cnn.com/2012/09/25/opinion/granderson-technology-phones/>
18. Kain, Helen. "Focus—Use It or Lose It." *MyAuthenticImpact Home Page.* 5 April 2016. <http://www.myauthenticimpact.com/2016/04/focus-use-it-or-lose-it/>
19. Pea, Roy, et al. "Media Use, Face-to-Face Communication, Media Multitasking, and Social Well-Being among 8–12 Year Old Girls." *Developmental Psychology* 48 (2012): 327–336.
20. Przybylski, A. K., and N. Weinstein. "Can You Connect With Me Now? How the Presence of Mobile Communication Technology Influences Face-to-Face Communication Quality." *Journal of Social and Personal Relationships,* 30 (2013): 237–246.

21 Piore, Adam. "What Technology Can't Teach Us about Happiness." *Nautil.us Home Page*. 17 September 2015. <http://nautil.us/issue/28/2050/what-technology-cant-change-about-happiness>

22 Bargh, John. "Automaticity in Cognition Motivation and Evaluation." *Yale University Home Page*. Accessed 17 August 2016. <http://www.yale.edu/acmelab/articles/Internet_and_Social_Life.pdf>

23 Attridge, Mark, Ellen Berscheid, and Jeffry A. Simpson. "Predicting Relationship Stability from Both Partners versus One." *Journal of Personality and Social Psychology* 69, no. 2 (1995): 254–68. doi:10.1037/0022-3514.69.2.254.

24 Hill, Charles T., Zick Rubin, and Letitia Anne Peplau. "Breakups before Marriage: The End of 103 Affairs." *Journal of Social Issues* 32, no. 1 (1976): 147–68. doi:10.1111/j.1540-4560.1976.tb02485.x.

25 Zickuhr, Kathryn, and Aaron Smith. "Digital Differences." Pew Research Center Home Page. 13 April 2012. <http://www.pewinternet.org/2012/04/13/digital-differences/>

26 *Alioscopy Home Page.* Accessed 15 April 2016. <http://www.alioscopy.com/en/principles.php>

27 Kushlev, Kostadin, and Elizabeth W. Dunn. "Checking Email Less Frequently Reduces Stress." Computers in Human Behavior. *Science Direct Home Page.* 1 February 2015. <http://www.sciencedirect.com/science/article/pii/S0747563214005810>

28 Ibid.

29 Gibson, Tom. "We Are Not Linear Processes." *Medium Home Page.* Accessed 31 July 16. <https://medium.com/better-humans/f0858a0cda88>

30 *KidsWifi Home Page.* Accessed 8 November 2016. <http:kidswifi.com>

31 Csikszentmihalyi, Mihaly. *Flow: The Psychology of Optimal Experience.* New York: Harper & Row, 1990.

Strategy #2: Know Thyself

1 Press Release. "IDC Forecasts Worldwide Shipments of Wearables to Surpass 200 Million in 2019, Driven by Strong Smartwatch Growth." *IDC Research Home Page.* 17 December 2015. <https://www.idc.com/getdoc.jsp?containerId=prUS40846515>

2 Furian, Peter Hermes. "Illusory Contours." *Shutterstock Home Page.* Accessed 7 November 2016. < http://www.shutterstock.com/pic-456101923.html>

3 Alba, Joseph W., and J. Wesley Hutchinson. "Knowledge Calibration: What Consumers Know and What They Think They Know." *Journal of Consumer Research* 27, no. 2 (2000): 123–56. doi:10.1086/314317.

4 McKeown, Les. "The Most Overlooked Key to Leading a Successful Company." *Inc.com Home Page* 22 October 2013. <http://www.inc.com/les-mckeown/secret-to-success-this-is-it.html>

5 Gladwell, Malcolm. *The Tipping Point: How Little Things Can Make a Big Difference*. Boston: Little, Brown, 2000.

6 Speech. Lorenz, E. N. (1972). "Predictability: Does the Flap of a Butterfly's Wings in Brazil Set Off a Tornado in Texas?" *139th Annual Meeting of the American Association for the Advancement of Science*. 29 December 1972.

7 Report. "State of the Global Workplace." Gallup Home Page. 2014. <http://www.gallup.com/services/178517/state-global-workplace.aspx>

8 "About the Quantified Self." *Quantified Self About Page.* Accessed 8 May 2015. <http://quantifiedself.com/about/>

9 "Lifelogging." *LifestreamBlog Home Page.* Accessed 17 May 2016. <http://lifestreamblog.com/lifelogging/>

10 Fox, Susannah, and Maeve Duggan. "Tracking for Health." Internet & Tech. *Pew Research Center Home Page*. 28 January 2013. <http://www.pewinternet.org/Reports/2013/Tracking-for-Health.aspx>

11 Kelly, Samantha Murphy. "The Most Connected Man Is You, Just a Few Years From Now: Does Chris Dancy Represent Our Enlightened Future or Lonley Self-Awareness?" *Mashable Home Page.* 21 August 2014. <http://mashable.com/2014/08/21/most-connected-man/>

12 *Feltron Home Page.* Accessed 8 May 2016. <http://feltron.com/info.html>

13 Ramirez, Ernesto. "Are They Improving with Data?" *Quantified Self Home Page*. 19 November 2014. <http://quantifiedself.com/2014/11/quantifying-classroom/>

14 Leonardi, Kevin. "Cerebral Curiosity: Graduate Student Steven Keating Takes a Problem-Solving Approach to His Brain Cancer." *Massachusetts Institute of Technology News Page*. 1 April 2015. <http://news.mit.edu/2015/student-profile-steven-keating-0401>

15 "The Human Cloud at Work: A Study into the Impact of Wearable Technologies in the Workplace." *Rackspace UK Home Page*. 1 April 2014. <https://www.rackspace.co.uk/sites/default/files/Human%20Cloud%20at%20Work.pdf>

16 Li, Ian. "Disasters in Personal Informatics: The Unpublished Stories of Failure and Lessons Learned." *Personal Informatics Home Page.* 14 September 2013. <http://www.personalinformatics.org>

17 Tierney, John. "Do You Suffer From Decision Fatigue?" *New York Times Magazine,* August 2011, MM33. <http://www.nytimes.com/2011/08/21/magazine/do-you-suffer-from-decision-fatigue.html?_r=0>

18 "Enjoy Work Again." *Spire Home Page.* Accessed 15 May 2015. <https://www.spire.io/enjoy-work>

19 Fogg, BJ. "Thoughts on Persuasive Technology." *Stanford Persuasive Tech Lab Home Page.* 1 November 2010. <http://captology.stanford.edu/resources/thoughts-on-persuasive-technology.html>

20 Byrnes, Nanette. "Technology and Persuasion." *MIT Technology Review Home Page.* 23 March 2015. <https://www.technologyreview.com/s/535826/technology-and-persuasion/>

21 "A Smart Coach by Your Side." *Jawbone Home Page*. 28 January 2015. <https://jawbone.com/blog/smart-coach-side/>

22 "NailO: Fingernails as an Input Surface." *MIT Media Lab Home Page.* Accessed 15 May 2015. <http://nailo.media.mit.edu/>

23 McFarland, Matt. "This Tattoo That Controls a Smartphone May Be a Glimpse of the Future." *CNN Money.* 15 August 2016. <http://money.cnn.com/2016/08/15/technology/mit-tattoo/>

24 Gomez, Jess. "Researchers Develop Smartphone Technology and App to Diagnose and Monitor Adrenal Gland Diseases." *Intermountain Healthcare Home Page*. 7 August 2014. <https://intermountainhealthcare.org/blogs/2014/08/smartphones-diagnose-adrenal-gland-diseases/>

25 Nguyen, My. "We Will Make You Sweat." *Wearable Technologies Home Page.* 17 February 2016. <http://www.wearable-technologies.com/2016/02/we-will-make-you-sweat/>

26 Gao, Wei, et al. "Fully Integrated Wearable Sensor Arrays for Multiplexed In Situ Perspiration Analysis." *Nature* 529, no. 7587 (2016): 509–14. doi:10.1038/nature16521.

27 Mann, Steve. "Wavelets and 'Chirplets': Time–Frequency 'Perspectives' with Applications." *Advances in Machine Vision Strategies and Applications* (1992), 99–128. doi:10.1142/9789814355841_0006.

28 Mann, Steve, et al. "Wearable Computing, 3D Aug* Reality, Photographic/Videographic Gesture Sensing, and Veillance." *Proceedings of the Ninth International Conference on Tangible, Embedded, and Embodied Interaction—TEI '14*, 2015. doi:10.1145/2677199.2683590.

29 Mann, Steve. "Phenomenal Augmented Reality: Advancing Technology for the Future of Humanity." *IEEE Consumer Electronics Magazine*, 4 (2015): 92-97. doi:10.1109/mce.2015.2463312.

30 *Meta Home Page.* Accessed 5 March 2015. <http://getameta.com>

Strategy #3: Train Your Brain

1 *Lifelong Kindergarten Home Page.* Accessed 5 March 2015. <https://llk.media.mit.edu>

2 Hernandez, Javier, Mohammed (Ehsan) Hoque, Will Drevo, and Rosalind W. Picard. "Mood Meter." Proceedings of the 2012 ACM Conference on Ubiquitous Computing—UbiComp '12, 2012. doi:10.1145/2370216.2370264.

3 ———. "Mood Meter: Counting Smiles in the Wild." *Proceedings of International Conference on Ubiquitous Computing (Ubicomp)*, September 2012.

4 Hernandez, Javier, Akane Sano, Miriam Zisook, Jean Deprey, Matthew Goodwin, and Rosalind W. Picard. "Analysis and Visualization of Longitudinal Physiological Data of Children with ASD." *Extended Abstract of IMFAR 2013*. San Sebastian, Spain: May 2–4, 2013.

5 Hernandez, Javier, Ian Riobo, Agata Rozga, Gregory D. Abowd, and Rosalind W. Picard. "Using Electrodermal Activity to Recognize Ease of Engagement in Children during Social Interactions." *International Conference on Ubiquitous Computing*. Seattle, WA: September 2014, 307–317.

6 Eknath, Easwaran. *Meditation: Commonsense Directions for an Uncommon Life*. Petaluma, CA: Nilgiri Press, 1978.

7 McGreevey, Sue. "Eight Weeks to a Better Brain: Meditation Study Shows Changes Associated with Awareness, Stress." *Harvard Gazette.* 21 January 2011. <http://news.harvard.edu/gazette/story/2011/01/eight-weeks-to-a-better-brain/>

8 Kaufman, Scott Barry. "Reasoning Training Increases Brain Connectivity Associated with High-Level Cognition." *Scientific American.* 18 March 2013. <http://blogs.scientificamerican.com/beautiful-minds/reasoning-training-increases-brain-connectivity-associated-with-high-level-cognition>

9 Malinowski, Peter, and Hui Jia Lim. "Mindfulness at Work: Positive Affect, Hope, and Optimism Mediate the Relationship between Dispositional Mindfulness, Work Engagement, and Well-Being." *Mindfulness* 6 (2015): 1250–62. doi:10.1007/s12671-015-0388-5.

10 Achor, Shawn. *The Happiness Advantage.* New York: Broadway Books, 2010.

11 Dweck, Carol S. *Mindset: The New Psychology of Success*. New York: Random House, 2006.

12 Fredrickson, Barbara L. "The Broaden-and-Build Theory of Positive Emotions." *The Science of Well-Being,* 2005, 216–39. doi:10.1093/acprof:oso/9780198567523.003.0008. <http://www.ncbi.nlm.nih.gov/pmc/articles/PMC1693418/pdf/15347528.pdf>

13 *Plasticity Labs Home Page.* Accessed 10 October 2016. <http://plasticitylabs.com>

14 *Lumo Lift Home Page*. Accessed 15 March 2015. <http://lumolift.com>

15 *Upright Home Page*. Accessed 15 March 2015. <http://upright.com>

16 Dweck, Carol. "Carol Dweck Revisits the 'Growth Mindset.'" *Education Week.* 22 September 2015. <http://www.edweek.org/ew/articles/2015/09/23/carol-dweck-revisits-the-growth-mindset.html>

17 Sterbenz, Christina. "I Tried a Startup That Claims to Make 86% of Users Happier . . ." *Business Insider.* 20 October 2016. <http://www.businessinsider.com/i-tried-a-startup-thats-supposed-to-make-me-86-happier-heres-how-it-works-2015-10>

18 Lyubomirsky, Sonja. *The How of Happiness: A Scientific Approach to Getting the Life You Want.* New York: Penguin, 2008.

19 Carpenter, Derrick. "The Science Behind Gratitude (and How It Can Change Your Life)." *Happify Daily.* 22 September 2015. <http://my.happify.com/hd/the-science-behind-gratitude>

20 Emmons, Robert A. *Thanks!: How Practicing Gratitude Can Make You Happier.* New York: Houghton Mifflin, 2008.

21 Alvaro Fernandez."How 'Saying Thanks' Will Make You Happier." *Huffington Post.* 17 November 2011. <http://www.huffingtonpost.com/alvaro-fernandez/how-saying-thanks-will-ma_b_76344.html?>

22 Steger, Michael F., and Todd B. Kashdan. "The Unbearable Lightness of Meaning: Well-Being and Unstable Meaning in Life." *The Journal of Positive Psychology* 8, no. 2 (2013): 103–15. doi:10.1080/17439760.2013.771208.

23 "The Science of Giving: Why Being Generous Is Good for You." *Happify Daily.* 15 March 2015. <http://my.happify.com/hd/science-of-giving-infographic/>

24 Barsade, Sigal G. "The Ripple Effect: Emotional Contagion and Its Influence on Group Behavior." *Administrative Science Quarterly,* 47.4 (2002): 644–75.

Strategy #4: Create a Habitat for Happiness

1 Bradley, Laura. "Snakes on the 'Glades." *US News and World Report.* 21 July 2014. <http://www.usnews.com/news/articles/2014/07/21/invasive-pythons-threaten-florida-everglades>

2 Shakespeare, William. *Hamlet.* Cyrus Hoy, ed. New York: W. W. Norton, 1996.

3 Waldrop, M. Mitchell. "The Chips Are Down for Moore's Law." *Nature.* 9 February 2016. <http://www.nature.com/news/the-chips-are-down-for-moore-s-law-1.19338>

4 Press Release. "IDC Forecasts Worldwide Shipments of Wearables to Surpass 200 Million in 2019, Driven by Strong Smartwatch Growth." *IDC Research Home Page.* 17 December 2015. <https://www.idc.com/getdoc.jsp?containerId=prUS40846515>

5 McGonigal, Kelly. "Why It's Hard to Let Go of Clutter." *Psychology Today Home Page.* 7 August 2012. <http://www.psychologytoday.com/blog/the-science-willpower/201208/why-it-s-hard-let-go-clutter>

6 Wolf, James, Hal R. Arkes, and Waleed Muhanna. "The Power of Touch: An Examination of the Effet of Duration of Physical Contact on the Valuation of Objects." *Judgment and Decision Making* 3, no. 6 August 2008, 476–82.

7 Hill, Graham. "Living with Less. A Lot Less." *New York Times*. 9 March 2013. <http://www.nytimes.com/2013/03/10/opinion/sunday/living-with-less-a-lot-less.html?pagewanted=2&_r=0&hp>

8 Doland, Erin. "Scientists Find Physical Clutter Negatively Affects Your Ability to Focus, Process Information." *Unclutterer Home Page*. 29 March 2011. <http://unclutterer.com/2011/03/29/scientists-find-physical-clutter-negatively-affects-your-ability-to-focus-process-information/>

9 Widrich, Leo. "What Multitasking Does to Our Brains." *Buffer App Home Page.* 26 June 2012. <http://blog.bufferapp.com/what-multitasking-does-to-our-brains>

10 Kondo, Marie. *The Life Changing Magic of Tidying Up: The Japanese Art of Decluttering and Organizing.* New York: Random House, 2014.

11 Wang, Yiran, Melissa Niiya, Gloria Mark, Stephanie M. Reich, and Mark Warschauer. "Coming of Age (Digitally)." Proceedings of the 18th ACM Conference on Computer Supported Cooperative Work & Social Computing—CSCW '15, 2015. doi:10.1145/2675133.2675271.

12 Ibid.

13 Green, Penelope. "Saying Yes to Mess." *New York Times.* 21 December 2006. <http://www.nytimes.com/2006/12/21/garden/21mess.html?pagewanted=all>

14 Baldé, C. P., F. Wang, R., Kuehr, and J. Huisman. "The Global E-waste Monitor—2014." *United Nations University, IAS—SCYCLE*. Bonn, Germany, 2015.

15 *Ship'N'Shred Home Page.* Accessed 8 November 2016. <http://shipnshred.com>

16 Lippe-McGraw, Jordi. "How to Feng Shui Your Digital Life in 20 Minutes or Less." *Today.com Home Page.* 4 November 2015. <http://www.today.com/health/how-feng-shui-your-digital-life-20-minutes-or-less-t53386>

17 Cho, Miakel. "How Clutter Affects You (And What You Can Do About It)." *Crew Home Page*. 24 January 2014. <https://ooomf.com/blog/how-clutter-effects-you-and-what-you-can-do-about-it/>

18 Becker, Joshua. "25 Areas of Digital Clutter to Minimalize." *Becoming Minimalist Home Page.* 11 June 2010. <http://www.becomingminimalist.com/25-areas-of-digital-clutter-to-minimalize/>

19 Lyubomirsky, Sonja. *The How of Happiness: A Scientific Approach to Getting the Life You Want*. New York: Penguin Press, 2008.

20 Benn, Caroline L., et al. "Environmental Enrichment Reduces Neuronal Intranuclear Inclusion Load but Has No Effect on Messenger RNA Expression in a Mouse Model of Huntington Disease." *Journal of Neuropathology & Experimental Neurology* 69, no. 8 (2010): 817–27. doi:10.1097/nen.0b013e3181ea167f.

21 Spires, T. L. "Environmental Enrichment Rescues Protein Deficits in a Mouse Model of Huntington's Disease, Indicating a Possible

Disease Mechanism." *Journal of Neuroscience* 24, no. 9 (2004): 2270–76. doi:10.1523/jneurosci.1658-03.2004.
22 Novak, Matt. "50 Years of the Jetsons: Why the Show Still Matters." *Smithsonian Magazine*. 19 September 2012. <http://www.smithsonianmag.com/history/50-years-of-the-jetsons-why-the-show-still-matters-43459669/>
23 Mackow, Rachel, and Jared Rosenbaum. "Closer Than We Think." 26 February 2011. <http://arthur-radebaugh.blogspot.com/p/closer-than-we-think.html>
24 Dudley-Nicholson, Jennifer. "The Jetsons' Vision of the Future 51 Years." *News Corp Australia Network*. 23 September 2013. <http://www.news.com.au/technology/the-jetsons8217-vision-of-the-future-51-years-ago-spoton/story-e6frfro0-1226725268775>
25 "Disney Previews New Mobile Games, New Ways to Experience Star Wars at 2016 Game Developer's Conference." *The Walt Disney Company Home Page*. 30 March 2016. <https://thewaltdisneycompany.com/disney-previews-new-mobile-games-new-ways-to-experience-star-wars-at-2016-game-developers-conference/>
26 Kolodny, Lora. "Why Apple Wants to Be the Smart Home's Nerve Center." *TechCrunch*. 13 June 2016. < http://techcrunch.com/2016/06/13/why-apple-wants-to-be-the-smart-homes-nerve-center/>
27 McCann, Laurenellen. "But What Is 'Civic'?" *Civic Hall*. 1 May 2015. <http://civichall.org/civicist/what-is-civic/>
28 Labarre, Suzanne. "Google London's New Office Is a Happy Kiddie Funhouse: It's Fingerpaintin' Time!" *FastCoDesign*. 27 January 2011. <http://www.fastcodesign.com/1663112/google-london-s-new-office-is-a-happy-kiddie-funhouse-slideshow>
29 Bock, Laszlo. "Two Minutes to Make you Happier at Work, in Life . . . and Over the Holidays." *LinkedIn Home Page*. 24 November 2014. <https://www.linkedin.com/pulse/20141124163631-24454816-two-minutes-to-make-you-happier-at-work-in-life-and-over-the-holidays?trk-SplashRedir=true&forceNoSplash=true>
30 Finch, Sidd. "Fancy Job Perks Won't Make You Happy." *The-Hustle Home Page*. 14 September 2015. <http://thehustle.co/fancy-job-perks-wont-make-you-happy>
31 Bock, Laszlo. "Google's Scientific Approach to Work-Life Balance (and Much More)." *Harvard Business Review*. 27 March 2014. <https://hbr.org/2014/03/googles-scientific-approach-to-work-life-balance-and-much-more>
32 Ibid.
33 2013 U.S. Workplace Survey. *Gensler.* 15 July 2013. <http://www.gensler.com/research-insight/research/the-2013-us-workplace-survey-1>

34 Thornton, Bill, Alyson Faires, Maija Robbins, and Eric Rollins. "The Mere Presence of a Cell Phone May Be Distracting." *Social Psychology* 45, no. 6 (2014): 479–88. doi:10.1027/1864-9335/a000216.

35 "CFO Survey Europe." Report. School for Business and Society, TIAS. Q2 2013. <http://www.cfosurvey.org/2016q2/Q2-2016-EuropeSummaryEnglish.pdf>

36 Lapowsky, Issie. "Get More Done," *Inc. Magazine,* April 2013.

37 Kaszniak, Alfred W., David M. Levy, Marilyn Ostergren, and Jacob O. Wobbrock. "The Effects of Mindfulness Meditation on Chronic Headaches, Stress, and Negative Emotions in High School Teachers." *Proceedings of Graphics Interface (GI '12).* Toronto, Ontario (28–30 May 2012). Toronto, Ontario: Canadian Information Processing Society, pp. 45–52.

38 "The Impact of Office Design on Business Performance." Report. *The Commission for Architecture & the Built Environment*, 2011.

39 Sargent, Kay. "Google Didn't 'Get It Wrong': A Deeper Look into That Recent WAPO Piece about Open Offices." *WorkDesign-Magazine.* 7 January 2015. <http://workdesign.com/2015/01/google-didnt-get-wrong-deeper-look-recent-wapo-piece-open-offices>

40 "What Workers Want." Report. *British Council for Offices*. 30 April 2013. < http://www.bco.org.uk/Research/Publications/What_Workers_Want2013.aspx>

41 Mayo, Keenan. "Five Broadway Lessons for Fixing Office Life." *Bloomberg Home Page.* 8 July 2013. <http://www.bloomberg.com/news/articles/2013-07-08/five-broadway-lessons-for-fixing-office-life>

42 Breene, Sophia, and Shana Lebowitz. "Why Are Google Employees So Damn Happy?" *Greatist Home Page.* 28 May 2013. < http://greatist.com/happiness/healthy-companies-google>

43 University of Exeter. "Designing Your Own Workspace Improves Health, Happiness and Productivity." *ScienceDaily.* 8 September 2010. <www.sciencedaily.com/releases/2010/09/100907104035.htm>

44 "Take a Two Minute Staycation in the Cigna Virtual Relaxation Pod." *ThisIsStory Home Page.* 2 February 2016. <http://thisisstory.com/take-a-two-minute-staycation-in-the-cigna-virtual-relaxation-pod/>

45 Randall, Tom. "The Smartest Building in the World." *Bloomberg Home Page.* 23 September 2015. <http://www.bloomberg.com/features/2015-the-edge-the-worlds-greenest-building/>

46 *Maki and Associates: Architects and Planning Home Page*. Accessed 16 August 2016. <http://www.maki-and-associates.co.jp/>

47 Tereshko, Elizabeth, and Zenovia Toloudi. "MIT Media Lab: Architecture as a Living Organism." *ShiftBoston Home Page.* 16 December 2011. <http://blog.shiftboston.org/2011/12/mit-media-lab-architecture-as-a-living-organism>

48 Cromwell, Sharon. "The School of the Future." *Education World Home Page.* 1998. <http://www.educationworld.com/a_curr/curr046.shtml#sthash.mplXzgO7.nD6SuHa9.dpuf>

49 *TaxiDog Educational Program Home Page*. Accessed 3 March 2016. <http://taxidogedu.org>

50 Madden, Mary, Amanda Lenhart, Sandra Cortesi, Urs Gasser, Maeve Duggan, Aaron Smith, and Meredith Beaton. "Teens, Social Media and Privacy." Report. *Pew Research Center.* 21 May 2013. <http://www.pewinternet.org/2013/05/21/teens-social-media-and-privacy/>

51 *KidsWifi Home Page.* Accessed 8 November 2016. <http://kidswifi.com>

52 Margalit, Liraz. "This Is What Screen Time Really Does to Kids' Brains." *Psychology Today* Home Page. 17 April 2016. <https://www.psychologytoday.com/blog/behind-online-behavior/201604/what-screen-time-really-does-kids-brains>

53 Chen, Brian X. "What's the Right Age for a Child to Get a Smartphone?" *The New York Times*. 20 July 2016. < http://www.nytimes.com/2016/07/21/technology/personaltech/whats-the-right-age-to-give-a-child-a-smartphone.html?em_pos=small&emc=edit_ml_20160721&nl=well-family&nl_art=4&nlid=74046621&ref=headline&te=1&_r=0>

54 *BreakFree Home Page.* Accessed 5 September 2016. <breakfree-app.com>

55 Perez, Sarah. "A New App Called Offtime Helps You Unplug without Missing Out." *TechCrunch Home Page.* 1 October 2014. <http://techcrunch.com/2014/10/01/a-new-app-called-offtime-helps-you-unplug-without-missing-out/?>

56 Przybylski, A. K., and N. Weinstein. "Can You Connect with Me Now? How the Presence of Mobile Communication Technology Influences Face-to-Face Conversation Quality." *Journal of Social and Personal Relationships* 30, no. 3 (2012): 237–46. doi:10.1177/0265407512453827.

57 Thornton, Bill, Alyson Faires, Maija Robbins, and Eric Rollins. "The Mere Presence of a Cell Phone May Be Distracting." *Social Psychology* 45, no. 6 (2014): 479–88. doi:10.1027/1864-9335/a000216.

58 Rhee, Hongjai, and Sudong Kim. "Effects of Breaks on Regaining Vitality at Work: An Empirical Comparison of 'Conventional' and 'Smart Phone' Breaks." *Computers in Human Behavior* 57 (2016): 160–67. doi:10.1016/j.chb.2015.11.056.

59 Matthews, Gail. "Goals Research Study." Research presented at the *Ninth Annual International Conference of the Psychology Research Unit of Athens Institute for Education and Research.* 28 May 2015. <http://www.dominican.edu/academics/ahss/undergraduate-programs/psych/faculty/assets-gail-matthews/researchsummary2.pdf>

Strategy # 5: Innovate Consciously

1 Weisberg, Jason. "We Are Hopelessly Hooked." *The New York Review of Books.* 25 February 2016. <http://www.nybooks.com/articles/2016/02/25/we-are-hopelessly-hooked/>

2 Calvo, Rafael A., and Dorian Peters. *Positive Computing: Technology for Well-being and Human Potential.* Cambridge: The MIT Press, 2014.

3 Helliwell, John F., Richard Layard, and Jeffrey Sachs, eds. *World Happiness Report 2012.* New York: Sustainable Development Solutions Network, 2013.

4 Tapscott, Don, and Anthony D. Williams. *Wikinomics: How Mass Collaboration Changes Everything.* New York: Portfolio, 2006.

5 Chivers, Tom. "The Story of Google Maps." *The Telegraph UK.* 4 June 2013. <http://www.telegraph.co.uk/technology/google/10090014/the-story-of-google-maps.html>

6 *Kickstarter About Page.* Accessed 1 November 2015. <https://www.kickstarter.com/about>

7 *Climate CoLab Home Page.* Accessed 1 November 2015. <http://climatecolab.org>

8 Evans, Philip, and Patrick Forth. "Borges' Map: Navigating a World of Digital Disruption." *Digital Disrupt Home Page.* 1 June 2015. <http://digitaldisrupt.bcgperspectives.com>

9 "The United States Digital Service." *The White House Home Page.* Accessed 1 August 2016. <https://www.whitehouse.gov/participate/united-states-digital-service>

10 Goldman, Jason. "Announcing South by South Lawn White House Festival of Ideas, Art and Action." *The White House Home Page.* 1 September 2016. <https://www.whitehouse.gov/blog/2016/09/01/announcing-south-south-lawn-white-house-festival-ideas-art-and-action>

11 Sheplelavy, Roxanne Patel. "Internet All Your Things." 20 July 2016. <http://thephiladelphiacitizen.org/internet-of-things-free-wifi-philadelphia/>

12 Sanders, Jennifer. *The Dallas Innovation Alliance Home Page.* 26 September 2016. <http://www.dallasinnovationalliance.com/news/2016/9/23/new-dallas-innovation-alliance-initiatives-highlighted-by-white-house>

13 Enwemeka, Zeninjor. "What's Boston's Score Today? City Launches Data Platform to Track Progress on Services." *WBUR News Home Page.* 15 January 2016. <https://www.wbur.org/2016/01/15/boston-cityscore-dashboard>

14 "CityScore." *City of Boston Home Page.* Updated 14 October 2016. <https://www.boston.gov/cityscore>

15 *The Better Block Foundation Home Page.* Accessed 14 October 2016. <http://betterblock.org>

16 Leson, Heather. “How Digital Humanitarians Are Closing Worldwide Disaster Response.” *Huffington Post.* <http://www.huffingtonpost.com/heather-leson/how-digital-humanitarians_b_9101950.html?utm_hp_ref=whats-working&utm_content=30768729&utm_medium=social&utm_source=facebook>

17 *Humanity Road Home Page*. Accessed 10 October 2016. <http://humanity-road.org/>

18 *Wine To Water Home Page.* Accessed 10 October 2016. < http://www.winetowater.org/>

19 Pritchard, Michael. “How to Make Filthy Water Drinkable.” *TED Home Page.* 1 August 2009. <https://www.ted.com/talks/michael_pritchard_invents_a_water_filter/transcript?language=en>

20 Connors, Mike. “Nansemond-Suffolk Student Takes a 3-D Printing Class on a Lark, Makes Prosthetics to Help Children.” *Virginian Pilot Home Page.* 19 August 2016. <http://pilotonline.com/news/local/education/nansemond-suffolk-student-takes-a--d-printing-class-on/article_badf3d56-19a1-5456-8704-5f81399e1837.html>

21 “Conscious Consumerism.” *The Center for a New American Dream Home Page.* Accessed 14 August 2016. <https://www.newdream.org/programs/beyond-consumerism/rethinking-stuff/conscious-consumerism>

22 Dunn, Elizabeth W., Lara B. Aknin, and Michael I. Norton. “Prosocial Spending and Happiness: Using Money to Benefit Others Pays Off.” *Current Directions in Psychological Science* 23: 41 (2014): 41–47. DOI: 10.1177/0963721413512503.

23 “Currents of Change: The KPMG Survey of Corporate Responsibility Reporting 2015.” Netherlands: Haymarket Netowrk Ltd., November 2015.

24 Dailey, Whitney. “Global Consumers Willing to Make Personal Sacrifices to Address Social Environmental Issucs.” *Cone Communications.* 27 May 2015. <http://www.conecomm.com/news-blog/2015-cone-ebiquity-csr-study-prcss-release>

25 Aknin, Lara B., Elizabeth W. Dunn, and Michael I. Norton. “Happiness Runs in a Circular Motion: Evidence for a Positive Feedback Loop between Prosocial Spending and Happiness.” *Journal of Happiness Studies* 13 (24 April 2011): 347–55.

26 “The GoodGuide Delivered to Your Phone.” *GoodGuide Home Page.* Accessed 20 August 2016. <http://www.goodguide.com/about/mobile>

27 *Better Business Bureau Home Page.* Accessed 20 May 2016. <http://www.bbb.org>

28 “The 2015 U.S. CSR RepTrak: CSR Reputation Leaders in the US.” *Reputation Institute.* 1 September 2015. <https://www.reputationinstitute.com/CMSPages/GetAzureFile.aspx?path=~%5Cmedia%5Cmedia%5Cdocuments%5C2015-us-csr-reptrak-report.pdf&hash=da63dd

6acee5a803dc962e9f093d17ce1f8f0f2d9823c9639274e2e7a3e2220d&ext=.pdf>

29 *B Corporations Home Page.* Accessed 10 September 2016. <http://www.bcorporation.net>

30 Rozenberg, Norman. "Tech a Driving Force in Socially Conscious Revolution." *Dell Home Page.* 30 June 2015. < http://www.techpageone.co.uk/downtime-uk-en/tech-driving-force-socially-conscious-revolution/>

31 Bryan, James H., and Mary Ann Test. "A Lady In Distress: The Flat Tire Experiment." ETS Research Bulletin Series 1966, no. 2 (1966): I–7. doi:10.1002/j.2333-8504.1966.tb00542.x.

32 Tsvetkov, Milena, and Micahel Macy. "The Science of 'Paying It Forward.'" *The New York Times Online.* 14 March 2014. <http://www.nytimes.com/2014/03/16/opinion/sunday/the-science-of-paying-it-forward.html?_r=1>

33 Dominus, Susan. "Is Giving the Secret to Getting Ahead?" *The New York Times Online.* 27 March 2013.< http://www.nytimes.com/2013/03/31/magazine/is-giving-the-secret-to-getting-ahead.html?_r=0>

34 Lin, Pei-Ying, Naomi Sparks Grewal, Christophe Morin, Walter D. Johnson, and Paul J. Zak. "Oxytocin Increases the Influence of Public Service Advertisements." *PLoS ONE* 8, no. 2 (2013). doi:10.1371/journal.pone.0056934.

35 Kramer, Adam D. I., Jamie. E. Guillory, and Jeffrey T. Hancock. "Experimental Evidence of Massive-Scale Emotional Contagion through Social Networks." *Proceedings of the National Academy of Sciences* 111, no. 24 (2014): 8788–90. doi:10.1073/pnas.1320040111.

36 Baldwin, Rashanah. "Yes, We've Got Crime, but Portrayals of Urban Neighborhoods as Hopeless Only Worsen the Problems." *The Reporters Home Page.* 1 April 2016. <http://www.thereporters.org/letter/yes-weve-got-crime/?utm_content=28117268&utm_medium=social&utm_source=facebook#more-1126 >

37 Obama, Barack. "President Obama: Why Now Is The Greatest Time to Be Alive." *WIRED,* 12 October 2016.

Glossary

1 Gilman, Hollie Russon. "The Future of Civic Technology." 20 April 2015. *Brookings.* https://www.brookings.edu/blog/techtank/2015/04/20/the-future-of-civic-technology/

2 Stone, Linda. "Continuous Partial Attention." *Linda Stone Home Page.* Accessed 24 October 2016. <https://lindastone.net/qa/continuous-partial-attention/>

3 *Digital Humanitarians Home Page.* Accessed 24 October 2016. <http://digitalhumanitarians.com>

4 Deans, Emily. “Dopamine Primer.” *Psychology Today Home Page.* 13 May 2011. <https://www.psychologytoday.com/blog/evolutionary-psychiatry/201105/dopamine-primer>

5 *Urban Dictionary Home Page.* Accessed 24 October 2016. <http://mindsetonline.com/whatisit/about/>

6 http://www.urbandictionary.com/define.php?term=fomo>

7 Burke, Brian. “Gartner Redefines Gamification.” 24 April 2014. *Gartner Blog.* <http://blogs.gartner.com/brian_burke/2014/04/04/gartner-redefines-gamification/>

8 Dweck, Carol. “What Is Mindset.” *Mindset Online Home Page.* Accessed 24 October 2016. <http://mindsetonline.com/whatisit/about/>

9 Achor, Shawn. *The Happiness Advantage: The Seven Principles of Positive Psychology That Fuel Success and Performance at Work*. New York: Broadway Books, 2010.

10 Bradley, Steven. “Design Principles: Visual Perception and the Principles of Gestalt.” *Smashing Magazine* Home Page. 28 March 2014. <https://www.smashingmagazine.com/2014/03/design-principles-visual-perception-and-the-principles-of-gestalt/>

11 Waldrop, M. Mitchell. “The Chips Are Down for Moore’s Law.” *Nature* 530, no. 7589 (2016): 144–47. doi:10.1038/530144a.

12 *Persuasive 2016* Home Page. Accessed 24 October 2016. <http://persuasive2016.org>

13 “Diminishing Returns | Economics | Britannica.com.” Accessed 24 October 2016. <https://www.britannica.com/topic/diminishing-returns.>

14 Eisenberg, Nancy, Richard A. Fabes, and Tracy L. Spinrad. “Prosocial Development.” *Handbook of Child Psychology* (2007). doi:10.1002/9780470147658.chpsy0311.

15 Wolf, Gary. “Know Thyself: Tracking Every Facet of Life, from Sleep to Mood to Pain, 24/7/365.” *Wired Home Page.* 22 June 2009. < https://www.wired.com/2009/06/lbnp-knowthyself/>

16 Howard, Ian P., and Brian J. Rogers. *Binocular Vision and Stereopsis*. New York: Oxford University Press, 1995.

17 Gielan, Michelle. *Broadcasting Happiness: The Science of Igniting and Sustaining Positive Change*. Dallas, TX: BenBella Books, 2015.

INDEX

B

C

D

E

F

G

H

I

J

K

L

M

N

O

T

ACKNOWLEDGMENTS

This book is a visual mosaic of years of thoughts, ideas, musings, and dreams shared by friends and family members in my life. I am eternally grateful to everyone who so selflessly poured love and energy into this project with me. Words alone cannot convey my gratitude, but they are at least a start.

Bobo—*Medofo*, you are my rock, my maven, my muse, and my best friend. This book wouldn't exist without you. Your ideas, encouragement, critiques, and edits have been invaluable. However, your unwavering belief in my potential and capabilities have meant more to me than anything else. When I started to lose my spark, you helped me find it again; when I lost my grounding, you brought me back to my third prong; and when I wanted to throw my laptop against the wall during the writing process, you gently reminded me to back it up first. Even though you were working on your own book, you selflessly stepped aside to make it possible for me to travel and write, all the while being "daddy of the year." Through it all, your love for me has shone through and words cannot express the depth of my gratitude.

Shawn—Thank you for pushing me off my bunk bed, physically and metaphorically. If you hadn't, I might never have realized that I was a unicorn. GoodThink might never have existed, and this book would never have been born. Thank you from the bottom of my heart for your encouragement of me as a speaker and author, but more than anything for your love for me as sister and co-adventurer in life.

Christiana, Gabriella, and Kobi Lyn—I wrote this book for you, so that the future may be brighter for all. But in the process of writing, you have given me so much in return. Christiana, thank you for dragging me into dance parties when I sat at my computer for too long. Gabriella, thank you for always greeting me with a big hug when I returned from a trip, whether I was gone for a few days at a talk or a few minutes at Starbucks; I always felt loved. Kobi Lyn, thank you for thoughtfulness and intuition, asking, "Mommy, are you happy?" when my mind got distracted, doubtful, or frustrated; you knew just how to pull me back to the present with your sweet voice and snuggles.

Mom and Dad—You raised Shawn and me as eternal optimists, despite the many challenges that you faced in your own lives. Thank you for the constant encouragement that we could be anything that we put our minds to and for always being there to support us when we launched our various endeavors. Dad, thank you for blowing our minds (both literally and figuratively) through your research in neuroscience and perception, and for letting us invade your labs on a regular basis (perhaps this is a good time to apologize for anything and everything we might have broken?). Mom, thank you for giving me a love of writing, and for the countless hours that you spent helping me learn how to edit and polish a story to make it come to life.

Michelle—From the first moment we met over a bowl of blueberries to the moment that I witnessed your vows as you married my brother, you have become a treasured part of my life. Your can-do spirit and vision for the world are an inspiration to me. Thank you for your continual encouragement, insight, and support throughout this book-writing process.

Chandra—Like a guardian angel, you entered my life at just the right moment and blessed me with your insight. Thank you for being my life coach and cheerleader, for holding me accountable and not letting me make excuses, and more than anything for giving me the nudge to embark on this path.

Jenny, Holli, and the team at Speakers Office—I can't imagine working with a better team. Your attention to detail, your thoughtfulness, and your support are unparalleled. When I decided to transition my role to speaking and writing, you didn't miss a beat but instead wholeheartedly embraced my journey and encouraged me on my way. Not only are you amazing partners but you have also become dear friends.

Glenn and the team at BenBella Books—Thank you for taking a chance on me as a new writer and for steering me toward this topic that has become my life work. I am so excited about bringing this research to life through writing, and am deeply grateful for the opportunity.

Debbie, Vy, and Brian—Thank you for your expert editorial help in bringing this book to life. Your attention to detail, structure, and content is incredible, and this book is as polished as it is because of you!

Jayme and the team at Worthy Media Group—Thank you for helping me get this book into the hands of readers around the globe so that I could share this research!

Alexis—Since the first time we met, you have always been in my corner and believed in me. We have shared many ups and downs in life, and through it all, you've always had an encouraging word or an idea to try. I love seeing your creativity come to life and look forward to seeing where your adventures take you as well!

Hannah—Thank you both for your design genius and support. I'm afraid to say too much more, lest others discover how amazing you are and whisk you away from me.

Jen—Thank you for your help putting together words, colors, and images to find my vibe. In the midst of launching your own brand, you generously shared your time and talent to breathe life into this project so I could share my vision and research with others.

Megan—My Starbucks writing buddy, do you know how clutch you were to my writing process? Day after day you sat with me to help me ward off the procrastination bug, and you inspired me with your own dedication to the writing process. I cannot wait to read *your* book in print in the very near future!

Janne—My unexpected friend, you came into my life and blessed everyone I came into contact with. Your selfless, giving spirit wows me on a daily basis and your authenticity inspires me. You were never afraid to stand in the gap when I was struggling, always checking in right when I started to doubt myself and helping me push through the mental hurdles. Thank you for the gift you have been to my whole family and me.

Kelci and Jordan—Thank you for inspiring me to overcome the "shoulds" and "musts" in life so that I could find my voice and create freely. Your love and light in my life spurs me onward, and your friendship to me and my whole family is such a gift.

Cathy and Greg—Our lives have run in parallel for so many years, and you have been there for each step along the way. We have laughed over tween drama, cried over unexpected life events, vented over frustrations, and shared many special memories. Thank you for enriching my life and that of my family as well.

My Bootcamp Girls—Thank you for inspiring me to (making me?) run a half marathon that helped me realize I could accomplish more than I ever dreamed. Thank you for running the distance with me throughout this book-writing process (and Cathy, I'm still sorry I sprinted to the finish without you).

Stephanie and Emily—For your patience and love, even when I disappeared for long stretches in the writing process, and for not giving up on me through it all. Your friendship means the world to me.

The Blankson family—For being my travel angels, always watching out for me, listening to me, praying for me, and encouraging me.

The Hammond family—For taking me on my first digital detox to Lake Shasta, for inspiring me with your own creativity, and for jumping in to help read and edit with a genuine enthusiasm that made me feel so loved.

All my friends and followers—For sharing your stories, ideas, apps, and more. Keep 'em coming!

ABOUT THE AUTHOR

Amy Blankson has become one of the world's leading experts on the connection between positive psychology and technology. She is the only person to have been named a Point of Light by two presidents (President George H. W. Bush and President Bill Clinton) for creating a movement to activate positive culture change. A sought-after speaker and consultant, Amy has now worked with organizations like Google, NASA, the US Army, and the XPRIZE Foundation to help foster a sense of well-being in the Digital Era. Amy received her BA from Harvard and MBA from Yale School of Management. Most recently, she was a featured professor in Oprah's Happiness course. Amy is the author of an award-winning children's book, *Ripple's Effect*, and is the mother of three girls who remind her on a daily basis why it is so important to create a happier future for all.